Investigation of Forrestal Fire

Copy of findings, recommendations and opinions of investigation into fire on board USS Forrestal (CVA 59)

By Chief of Naval Operations 1969

Algorithmically Annotated by Nimble Books LLC 2022

Algorithmic Summaries

Annotations provided by Nimble Books LLC.

Methods

–Recursive, transecting, abstracting summary

–Reduced 37,446 words to 629 words (1.6%)

–Rounds of recursion: 3, chunk length: 2000

–Machine-generated; known to contain errors, use with caution

One-word TL;DR

Negligence.

TL;DR

The Board of Investigation into the fire on board the USS FORRESTAL found that poor and outdated doctrinal and technical education, as well as deficient supervision, were contributing causes. The commanding officer of the USS FORRESTAL is responsible for the fire that occurred on the ship on July 29, 1967, which killed and injured many people.

Action Items

1. Evaluate the effectiveness of ordnance safety procedures and make improvements as necessary.

2. Develop new firefighting equipment and procedures specifically for use on aircraft carriers.

3. Pre-position firefighting resources near aircraft carriers to ensure rapid response in the event of a fire.

4. Modify existing carrier firefighting systems to improve their effectiveness in responding to fires.

Final Round of Summaries (7 Chunks)

A fire on board the USS Forrestal in 1967 resulted in the deaths of 134 crew members and injuries to 161 others. The fire was started by a chain of events that began when a rocket ignited a fire on the ship's flight deck, causing several explosions.

The fire on the USS Forrestal was caused by an electrical malfunction, and quickly spread due to the presence of flammable materials and the lack of effective fire-fighting measures. The high ambient temperature may have also contributed to the fire's spread.

The board of investigation into the 1967 fire on the USS Forrestal found that no disciplinary action was warranted against any personnel. They met with and assisted the preliminary investigation, which uncovered and committed certain key witnesses. The Board took testimony from witnesses and determined who should be designated parties on 6 August, with twenty persons being designated as parties. Such determinations were made with the following twenty persons designated as parties: B. USN, I USN, Jr., USN, 2 Commanding Officer, USS FORRESTAL (CVA-59) Commanding Officer, VF-11 Commander, Carrier Air Wing 17 L Since A02 USN USN, N USNR, USN, Jr., USN USN SN, USN USN USN, USN ISN, and AOAA SN USN USN USNR 3 Air Officer, USS FORRESTAL (CVA-59) All redactions are B-6 Engineering Officer, USS FORRESTAL (CVA-59) Carrier Air Wing 17 Senior ISO and pilot of aircraft #110 on 29 July 1967 Hangar Deck Officer, USS FORRES TAL (CVA-59) Damage Control Assistant, USS FORRESTAL (CVA-59) VF-11 Maintenance Officer Carrier Air Wing 17 Ordnance Officer

Flight Deck Officer, USS FORRESTAL (CVA-59) VF-11 Div. Off. Avionics/Weapons Div. VF-11 Weapons Branch Officer Fire Marshall, USS FORRESTAL (CVA-59) Air Gunner, USS FORRESTIL (CVA-59) VF-11 Assistant Avionics/ Weapons Div. Off. VF-11 Weapons Branch Chief VF-11 Leader of an ordnance team were patients at the Naval Hospital, Subic Bay, R. . counsel for the Board informed them of their designation as parties at the hospital on the night of 6 August. Their rights were also explained to them at that time and they both signed statements acknowledging the explanation of their rights. (Enclosure (2)). VF-11 Flugged-in Rockets on aircraft #110 VF-11 Conducted stray voltage test on aircraft #110 Neither of these parties desired counsel at that time. The next morning, 7 August 1967, the remaining 18 parties were informed of their designation by the Board. Their rights were explained to them by counsel and they all signed statements acknowledging that their rights had been explained to

An investigation board was set up to look into a fire on the USS Forrestal. The board interviewed many witnesses, conducted experiments, and used the Flat tape as evidence. The board resolved conflicts in testimonies and found that no disciplinary action is necessary.

The 1967 accident in which an F-4B aircraft exploded during a stray voltage check was due to a lack of documentation specifying when and how the check should be conducted, so the check was conducted while the aircraft engines were being started, which caused the explosion. 134 people were killed in the fire that resulted.

The Weapons Coordination Board met on 29 June 1967 to establish ordnance handling procedures for the upcoming WESTPAC deployment. The procedures were validated by inspectors during an Operational Readiness Inspection. The minutes from the meeting were given to squadrons for correction of ambiguous statements.

This document contains findings of fact regarding an issue with the wiring of a safety switch in triple ejector racks (TERs) in use aboard the USS FORRESTAL. The incorrect wiring made it possible for a high magnitude voltage to be applied to the rocket motors, causing them to fire unexpectedly.

IN REPLY REFER TO
Op-0982/ipl
Ser 09P098

2 1 AUG 1969

– Unclassified when enclosure (1) is removed

From: Chief of Naval Operations
To: Judge Advocate General

Subj: Investigation of Forrestal Fire; return of

Ref: (a) JAG ltr ser 0132 of 12 August 1969

Encl: (1) Copy of findings, recommendations and opinions of investigations
into fire on board USS Forrestal (CVA 59) and First, Second
and Third endorsements thereon

1. As requested in reference (a) enclosure (1) is returned.

B-6

By direction

691209-0041

THIRD ENDORSEMENT on ___ CSN, 070031, ltr of 19 Sep 1967

From: Judge Advocate General
To: Chief of Naval Operations
Via: (1) Commander, Naval Air Systems Command
    (2) Commander, Naval Ships Systems Command
    (3) Commander, Naval Ordnance Systems Command
    (4) Chief of Naval Personnel (P-21)

Subj: Inves. - Fire and explosions aboard USS FORRESTAL (CVA 59)
on 29 July 1967; deaths of 116 persons; presumptive deaths
of 18 persons (missing or remains incapable of positive
identification); injuries to 161 personnel; ordered by
Commander Naval Air Force, U. S. Atlantic Fleet,
30 July 1967 (C)

Encl: (1) Record of subject investigation with copies of First
and Second Endorsements
(2) CNO ltr, Op-09B1L/jm, Ser 0859P09B1L, dtd 25 October 1968,
with enclosures

1. (U) Forwarded. All deaths, verified and presumptive, and all
injuries reported in enclosure (1) occurred in the line of duty and
not as the result of any misconduct on the part of the individuals
killed, missing, and injured.

2. (C) The catastrophic fire aboard the FORRESTAL resulted from a
chain of events starting at about 1051 local time (Hotel) on
29 July 1967. The FORRESTAL had been on Yankee Station in the Gulf
of Tonkin since 24 July, and had been launching air strikes against
North Vietnam since 25 July. Other carriers in the area were the
ORISKANY and BON HOMME RICHARD; the INTREPID was en route Yankee
Station; the CONSTELLATION was moored in Subic Bay. Captain
B-6 , USN, was commanding FORRESTAL. Rear Admiral B-6
USN, was commanding the Task Group comprised of FORRESTAL,
and Destroyers RUPERTUS and TUCKER. Commander B-6
USN, commanded Carrier Air Wing SEVENTEEN, aboard the FORRESTAL.
Commander B-6 , USN, commanded Fighter Squadron ELEVEN,
which was assigned to Air Wing SEVENTEEN at the time.

220409

014 709-0241

3. (C) 12 A-4, 12 F-4, and 3 A-6 aircraft were on the deck ready for a launch to begin at about 1100. Each airplane was variously loaded with bombs, missiles, rockets, and 20 mm. ammunition. A preliminary launch of one KA-3B and one EA-1F had been made at 1050; launches of another KA-3B and one E-2A were in progress. Three RA5C aircraft were on the starboard side of the flight deck, abaft the island. A-4E #405, piloted by Lieutenant Commander ██████, USN, was the third aircraft forward of the stern on the port side of the flight deck. An F-4, #110 was spotted on the extreme starboard quarter of the flight deck, headed inboard at approximately a 45° angle to the ship's head. Lieutenant Commander ██████, USN, was pilot; Lieutenant (junior grade) ██████ was preflighting the rear cockpit; eight crewmen were engaged in checking ordnance and assisting in starting and otherwise readying the plane for launch. External electrical power was being supplied to the airplane in connection with starting the starboard engine. Surmisedly due to a combination of material deficiencies and team operational procedures affording less than maximal emphasis upon safety, in the course of switching from externally supplied electrical power to the internal power system of the F-4, sufficient electrical current reached one of three launchers on the port inboard wing station to fire a ZUNI rocket. The rocket crossed the flight deck and struck A-4 #405, some 100 feet away, rupturing the full 400-gallon fuel tank on the A-4 and igniting the jet fuel. A fragment punctured the centerline external fuel tank of another A-4 just aft of the jet blast deflector of catapult #3. Fuel from this tank poured on deck and was ignited. The burning fuel spread aft, fanned by 32 knots of wind from 350° relative and by exhausts of at least three jets forward.

4. (C) Fire quarters, then general quarters, were sounded at 1052 and 1053. An AN-M65 1000-pound bomb fell to the deck from A-4 #405, came to rest in a pool of burning jet fuel, split, and was observed to be burning brightly. Within one and one-third minutes after initiation of the fire, the first hose began to play salt water on the forward boundary of the fire. Some fourteen seconds later, a bomb exploded on the flight deck with 35 personnel in close proximity. This explosion decimated two hose teams and caused 27 other casualties, plus spreading the fire to three A-4 aircraft spotted across the stern. Nine seconds later a second bomb exploded at the after end of the flight deck, even more violently than the first, hurling bodies and debris as far as the bow. The second major explosion extended the fire along 7 F-4's and toward the 3 RA5C's on the starboard side abaft

the island. Further, the second explosion interrupted effective
fire fighting efforts on the flight deck. Five other major
explosions followed, precluding resumption of effective fire
fighting for some five minutes. During that five minutes, however,
jettisoning of ordnance, care of the injured, and preparations for
further fire fighting were carried on in the shelter of the island.
Some 40,000 gallons of jet fuel aboard burning aircraft on the
flight deck fed the flames. Fuel flowed over the sides and stern,
setting fires on the sides, sponsons, fantail, and in hangar bay 3.
The force of bombs exploding on the flight deck penetrated to
hangar bay 3, starting fires on the 03, 02, and 01 decks aft. The
force of the explosions killed some fifty sleeping night check crew
personnel, and others, for a total of 91 killed in the after areas
of the ship.

5. (C) The respite following the seven major explosions permitted
concentrated damage control efforts, including moving unaffected
aircraft forward, jettisoning aircraft that were burning or leaking
fuel, establishing fire boundaries, and fighting fire progressively
aft. Fire main pressure was maintained at 150 pounds per square
inch throughout. Flames on the flight deck were extinguished by
1140, although smouldering continued for some time thereafter.
The last fire on the ship was extinguished at 0400 the following
morning.

6. (C) One-hundred thirty-four personnel were dead or missing as
a result of the fire and explosions; 161 were injured. Damage to
the ship - exclusive of aircraft and air equipment damage - was
estimated as of 15 September 1967 as being in excess of $72 million.

7. (C) The Investigating Officers were of the opinion that poor
and out-dated doctrinal and technical documentation of ordnance and
aircraft equipment and procedures, evident at all levels of command,
was a contributing cause of the accidental rocket firing which was the
first event in the catastrophic chain. It was recognized that these
were high-tempo combat operations, but the investigation considered
that approved procedures should have been followed nonetheless with
any modifications required to fit combat operational requirements
being obtained officially.

8. (C) The Investigating Officers were of the opinion that no blame
attached to the Commanding Officer of the FORRESTAL or any of the

officers of his command; to the Commander of Carrier Air Wing
SEVENTEEN; or to the pilot of the plane from which the ZUNI
rocket was fired. The Commanding Officer of Fighter Squadron
ELEVEN was deemed to have demonstrated poor judgment and
supervisory deficiencies in not maintaining clearly prescribed
and consistently adhered to procedures, although no specific
blame in connection with the fire was attributed to him.
Lieutenant (junior grade) ~~~~~~~~ B-6 ~~~~~~~~, USNR, Weapons
Branch Officer, Fighter Squadron ELEVEN, was deemed deficient
in the degree of supervision and control exercised over personnel
in the ordnance branch, but no blame was deemed to attach to him
in connection with the fire. The investigation recommended that
no disciplinary or administrative action be taken with regard to
any persons attached to the FORRESTAL or Carrier Air Wing
SEVENTEEN as a result of the fire.

9. (C) Following the fire the FORRESTAL was brought to Naval
Air Station, Cubi Point, Republic of the Philippines. The
Investigating Officers arrived on board on 3 August 1967 and
began proceedings on that date. The investigation continued
during some 32 days that the FORRESTAL was engaged in making a
transit from the Philippines to the United States.

10. (C) Under date of 26 September 1967, the Commander, Naval
Air Force, U. S. Atlantic Fleet, who ordered the investigation,
generally approved the findings, opinions, and recommendations.
Information copies were forwarded to CINCPAC, NAVAIRPAC,
CONSEVENTHFLT, and CTF77. Copies of VOLUME 1, containing
findings, opinions, and recommendations, were transmitted to
all Atlantic Fleet Carrier Division Commanders.

11. (C) The Commander in Chief, U. S. Atlantic Fleet, under
date of 1 December 1967, endorsed the record noting that a
cause sine qua non of the casualty was probably the connection
of the rocket harness - commonly called the pigtail, rocket
cable, or rocket pigtail - the cable making the electrical
connection between the Triple Ejector Rack (TER) and the
launcher for the rocket - at an early stage of the pre-launch
plane preparation, rather than waiting until just before the
plane was to be launched from the catapult. NAVWEPS Op 3347
was quoted in pertinent part: "The pigtail connector shall not
be plugged into the launcher receptacle until just before takeoff."

CONFIDENTIAL

Commander I B6 were promoted during the interim.

13. (U) The investigation   based therein
appear to have been conducted in accordance with pertinent
regulations contained in the JAG Manual. The record is routed for
information and action within the cognizance of departmental
offices addressed, and for ultimate return to this Office to be
filed.

B6

Acting Judge Advocate General

~~CONFIDENTIAL~~
~~CONFIDENTIAL~~

Ser 01606 /14
1 December 1967

CONFIDENTIAL

SECOND ENDORSEMENT on ~~B-6~~   USN, 070031/1310,
ltr of 19 September 1967

From: Commander in Chief U. S. Atlantic Fleet
To: Judge Advocate General

Subj: USS FORRESTAL (CVA 59) Fire Investigation (U)

Encl: (406)

      (407)

      (408)

      (409)

1. (U) Forwarded.

2. (C) A review of the voluminous material contained in the Report of Investigation establishes the central fact that a ZUNI rocket was inadvertently fired from an F-4 aircraft (#110) and struck the external fuel tank of an A-4 aircraft (#405) which was clustered in a pack along with other aircraft on the flight deck of the USS FORRESTAL. This inadvertent firing of the rocket resulted in a raging fire. Additionally, although the Report of Investigation contains a considerable amount of information about stray currents, switches, shorting devices, safety pins, external and internal power in the aircraft, it is highly improbable that the rocket would have fired with the resulting damage had the pigtails not been connected to the launcher receptacle until the aircraft was ready for takeoff, as required by pertinent regulations. It is obvious from the testimony of at least two witnesses that the pigtails were connected to the launcher receptacle while the aircraft were clustered in the pack and prior to the starting of the aircraft and their moving to the catapult for takeoff. (Finding of Fact 142).

3. (C) It may be argued that NAVWEPS OP 3347 which states in

DOWNGRADED AT 3 YEAR INTERVALS
DECLASSIFIED AFTER 12 YEARS
DOD DIR 5200.10   133   CONFIDENTIAL

~~CONFIDENTIAL~~

5.   (C) In view of the foregoing, and in view of Findings of
Fact 237, 239, 240, 243, 244, 245, 246, 247, 248, 250, 252,
262, and 263, it is clearly established that the standard
operating procedures of VF-11 were vague and undocumented
and command attention from the Commander, Carrier Air Wing

~~CONFIDENTIAL~~

- pertinent part that "The pigtail connector shall not be plugged
into the launcher receptacle until just before takeoff", and
NAVWEPS OP 2210 Volume I which states in pertinent part, "Arm
the launcher just before aircraft takeoff", are subject to
interpretation and that perhaps the ordnance personnel com-
plied with the safety regulations. However, it is an irrefut-
able fact that the regulations are precautionary and should
therefore in all circumstances be interpreted in favor of
safety. Further, the sensitivity of the ZUNI rocket and the
launching device is well known and the possibility of inadvertent firing of other ordnance should be particularly
in the minds of personnel charged with responsibilities in
connection with the loading and arming of ordnance on combat
aircraft. It is also considered that ordnance personnel are,
or should be, well versed in pertinent safety regulations
issued in connection with the types of ordnance being handled
at any particular time. This fact alone makes a decision to
intentionally fail to comply with two (2) published safety
regulations all the more incredible.

4.   (C) The Weapons Coordination Board in its memorandum dated
29 June 1967 stated with regard to NAVWEPS OP 2210 which for-
bids the arming of launcher armament until "just prior to
takeoff" to have been traditionally interpreted on carriers
to mean "on the catapult". Yet, the "Final Recommendation"
of the Board was to "Allow connection of rocket pack pigtail
prior to aircraft taxi to catapult". Further in this connection, the Report of Investigation establishes the fact
that LT   B b   , USN, the Ordnance Officer for
Carrier Air Wing SEVENTEEN, and LTJG   B c
USNR, the Weapons Officer, VF-11, each briefed his superior
with regard to the meeting and the "Final Recommendation".
It is also noted that the draft memorandum of the Weapons
Coordination Board was not, at any time, delivered to or seen
by the Commanding Officer, CVA 59, and was not, in fact, dis-
seminated outside of the Operations Department of the ship.
(Finding of Fact 229). The Commander in Chief U. S. Atlantic
Fleet concurs specifically in Opinions 17 and 18 wherein
procedures contrary to those set forth in NAVWEPS OP 3347 and
NAVWEPS OP 2210 cannot be amended or superseded except by the
issuing authority, in this case, the Naval Weapons System Command
This fact prevails even under high tempo combat operations.

~~CONFIDENTIAL~~

7.   (C) Further in this connection, the Report of Investigation
establishes the fact that CAP' B. b   JSN, was the
Commanding Officer of the USS FORRESTAL on the date of the
fire having assumed command on 7 May 1966. Although it was
the opinion of the Board of Investigation, "That no improper
acts of commission or omission by personnel embarked in
FORRESTAL directly contributed to the inadvertent firing of
the ZUNI rocket from F-4 #110" (Opinion #4) and "That the
deaths and injuries resulting from the fire aboard FORRESTAL
on 29 July 1967 were not caused by the intent, fault, negligence,
or inefficiencies of any person or persons embarked in FORRESTAL"
(Opinion #115), it is considered that the conduct of the
Commanding Officer and the role played by him in connection with
the tragic incident cannot be ignored. Assessment of the
appropriate degree of personal responsibility of the Commanding
Officer of a ship as large and complex as a modern aircraft
carrier for the results of a casualty occurring in but one area
of the ship's activities is most difficult. The Commanding
Officer, in a broad sense, is responsible for everything that
goes on in his ship. Yet, the myriad of activities, the thousands
of people, and the vast array of equipment and material that
constitute an operating carrier, necessarily limit the direct
personal attention which the Commanding Officer can give to
each activity. Accordingly, his direction of the activities
of the ship is exercised through the issuance of orders,
preparation of directives, review of procedures, inspections,
and by interviews and counselling of his subordinates. When
considering an area as important and potentially dangerous as
the handling of rockets, particularly the ZUNI rockets, which
are known throughout the Naval Establishment to have inadvertently fired in the past, it is considered only reasonable that
a Commanding Officer would have been diligent in showing
particular interest in the procedures utilized in loading the
weapons aboard an aircraft and the arming of those weapons.
On 12 June 1967, the Commanding Officer issued regulations
(USS FORRESTAL Instruction 8020.2) which required "POSITIVE
COMPLIANCE" with NAVWEPS OP 3347 and NAVWEPS OP 2210.
Yet, the Commanding Officer so far as the record indicates did

CONFIDENTIAL

not make inquiry concerning compliance with applicable safety
regulations and alleges that he was not informed by any of the
ship's company. This fact has special significance in light of
two factors: (a) the danger involved when less than absolute
compliance with safety precautions is authorized or tacitly
approved, and (b) it is generally known in the naval aviation
community that the activity of the aircraft carrier over which
the Commanding Officer can best exercise his leadership, his
supervision and enforce his own personal standards on the performance of his subordinates is that on the flight deck of the
ship, in this case, where the tragedy occurred.

8.   (C) The degree of responsibility of the Commanding Officer
for the inadvertent firing of the rocket does not lie in the
magnitude of the disaster, but rather in an assessment of his
execution of responsibility for proper administration and operation
of the ship and its Weapons Department overall. Although he must
have been aware of NAVWEPS OP 3347 and NAVWEPS OP 2210 and their
safety regulations pertinent to handling of aircraft rockets,
since each was referred to in the FORRESTAL Instruction 8020.2,
the Commanding Officer directed the Executive Officer and other
officers of the ship to task the Weapons Planning Board to
develop ordnance handling regulations and procedures which met
safety requirements but also incorporated WESTPAC practices. The
recommendations of this Board embodied violations of safety
regulations promulgated in the NAVWEPS Ordnance Publications.
Since he knew also that there was a difference between Atlantic
Fleet training procedures and WESTPAC procedures, it would appear
that these facts alone should have alerted the Commanding Officer
for particular attention to this very area.

                                                  . It has
been administratively determined that Captain   b-b   is currently
assigned to the Office of the Chief of Naval Operations. By copy
of this endorsement and in accordance with section 0101a(3) of
reference (a), enclosure (409) is forwarded to the Chief of Naval
Operations for such action as is deemed to be appropriate.

9.   (C) It is noted that Section IV, Volume I of the Report of
Investigation does not reflect as a Finding of Fact that the In-
spector General, U. S. Atlantic Fleet conducted an inspection of
damage control readiness of the FORRESTAL on 10 May 1967. It has

CONFIDENTIAL

                       600 C3-0011

been administratively determined that a copy of the report was
made available to the Board. FORRESTAL was inspected as a
unit of Commander Naval Air Force, U. S. Atlantic Fleet in
determining the damage control readiness of the U. S. Atlantic
Fleet. The inspections of individual ships were made on short
notice and were designed to determine a ship's current capability
to maintain water tight integrity, fight fires, and repair
damages. As of the date of the inspection, 10 May 1967,
FORRESTAL was evaluated by INSGENLANTFLT as unsatisfactory in
damage control readiness. Opinion 54, Page 112 of the Report
of the Board indicates that the FORRESTAL's material readiness
for fire fighting and damage control was at acceptable standards
at the time of the fire.

10. (C) Further in this regard, the Report reveals that a
substantial number of Air Wing personnel were inadequately
trained in the fundamentals of ship's damage control organization and operation. (Paragraphs 311, 316, 326, 330, and 337).
The INSGENLANTFLT's inspection aboard the FORRESTAL on 10 May
1967 revealed that the repair parties above the second deck were
led by aviation ratings who displayed a lack of damage control
organization and knowledge. The report does not indicate whether
or not this condition still prevailed on 29 July 1967. Additionally
the Report does not list as a Finding of Fact any attempts to
maneuver the ship. Further, there is no stated opinion of the
Board as to whether maneuvering the ship such as making tight turns
to sluice fuel overboard by heeling the ship would have helped or
hindered fighting the fire and jettisoning aircraft or ordnance.

11. (C) With respect to Opinion 53 and Recommendation 30, the
Commander in Chief U. S. Atlantic Fleet is, and has been, concerned with the high turnover rate of enlisted personnel in the
Fleet units. This turbulence is caused by a combination of
factors, including the low reenlistment rate as mentioned by
COMNAVAIRLANT. Other causes are low manning in supervisory
pay grades (E5 to E9), which as of 31 August 1967 was approximately 82% of allowance, inter-Fleet transfers of personnel to
support Southeast Asia Operations, an increase in the tempo of
Fleet operations, and the commissioning and reactivation of
Fleet units. All of these causes place a heavy drain on the

CONFIDENTIAL

personnel resources available to the Fleet and Type Commanders.
The proposal of COMNAVAIRLANT to conduct test manning in the
USS AMERICA requires additional amplification and supporting
data. Accordingly, COMNAVAIRLANT is requested to provide Commander in Chief U. S. Atlantic Fleet additional information in
order to further consider this recommendation.

12.   (C) The Commander in Chief U. S. Atlantic Fleet concurs
in the statements contained in paragraph 4 of the First Endorsement. It is recommended that the feasibility of designing a
single test set that would enable stray voltage testing of A.C.,
D.C., and R.F. voltage at the attachment point of the rocket
harness be investigated. Such a test set would provide a comprehensive test of the circuitry and limit the possibility of
attaching the rocket harness to the LAU 10/A Launchers with
any form of stray voltage present in the circuit.

13.   (C) In regard to Finding of Fact #293, which states that
there are three (3) different ordnance safety pins in service
in the Navy that will fit the TER rack, it is recommended that
a study be initiated to determine the feasibility of designing
a single size safety pin for all Navy aircraft that would render
all components of a weapon system safe. Such a pin could be
utilized without regard to insuring that the correct pin is inserted into the correct component. This type of safetying method
would preclude the possibility of inserting a safety pin into a
component and not actually rendering the component safe. It is
realized that such a system may not prove to be a feasible retrof
for current operational weapon systems, however, it is felt that
such a system would be feasible if included in a design specifi-
cation during development of future weapon systems.

14.

15.   (C) The Board of Investigation affixes responsibility for
the deaths and injuries resulting from the fire aboard
FORRESTAL on 29 July 1967

B-5

(Opinion #116).

view of the remarks herein before, CINCLANTFLT does not agree wit
this opinion. Certain instructions and technical publications ma
have been lacking or inadequate, for which the Air Systems Comman
may be at fault, but the degree to which this may have contribute
to this fire cannot be ascertained, much less can it be ascribed
as the only cause for the fire.

139

CONFIDENTIAL

~~CONFIDENTIAL~~

FIRST ENDORSEMENT on ฿-6 'N ltr of 19 September 1967

From: Commander Naval Air Force, U.S. Atlantic Fleet
To: ~~Judge Adv~~

UNCLASSIFIED

Ser 01606 /14
1 December 1967

~~CONFIDENTIAL~~

15. (U) Subject to the foregoing, the proceedings, findings of fact, opinions and recommendations, as approved by the First Endorsement, are approved.

฿-6

Copy to:
CNO
CINCPACFLT
COMNAVAIRLANT
COMNAVAIRPAC
COMSEVENTHFLT
NAVAIRSYSCOMHQ
CTF 77
COMCARDIV TWO
COMCARDIV FOUR
COMCARDIV SIX
COMCARDIV FOURTEEN
COMCARDIV SIXTEEN
COMCARDIV TWENTY

690 11

UNCLASSIFIED

~~CONFIDENTIAL~~

Copy

Volume I Enclosure (7) to CNO Letter ser .... of 10-25-68

7360-67

CONFIDENTIAL

Subj: USS FORRESTAL (CVA-59) Fire Investigation (U)

6. Certain increases in allowances of OBA's and OBA cannisters can be accomplished. However, prior to a decision on adoption of recommendation 38, detailed value analysis must be applied to the proposition, including preparation of space layout for stowage of added allowances and engineering estimates of cost to provide stowage facilities. Additionally, in the event the decision is made to proceed with the maximum contemplated in the recommendation, implementation should be delayed until line item budgeting and appropriation is accomplished.

7. Recommendation 47 is concurred in. In addition to the labelling of each item of ordnance with cook-off times, it is recommended that a placard be published similar to tables 1 and 2 of Special Weapons Ordnance pamphlet (SWOP) 20-11, giving withdrawal distance, fire fighting/withdrawal time and explosive ignition time. This placard should be readily available at control points such as Flight Deck Control, Primary-fly, the Bridge, Damage Control Central, Central Control and Aviation Ordnance Control Station.

8. Subject to the above, the proceedings, findings of fact, opinions and recommendations of the board of investigation are approved.

9. The board of investigation has exhibited great resourcefulness and ingenuity in its dedication to seek out and to develope every possible source of information touching on the causes of this disaster. All facts and suppositions were explored to determine whether or not they were authentic and provable and, more important, whether they were useful in preventing a similar occurrence in the future. Inaccuracies, including confused and conflicting testimony, have been eliminated. Immediate and future preventative measures have been recommended. This thoroughness on the part of the members and counsel of the board has resulted in a comprehensive report which is of inestimable value to the naval service. By separate correspondence, appropriate commendatory action will be instituted by Commander Naval Air Force, U.S. Atlantic Fleet.

B-6

Copy to:
CINCPACFLT (complete, less plat film)
COMNAVAIRPAC (complete, less plat film)
COMSEVENTHFLT (complete, less plat film)
CTF 77 (complete, less plat film)
COMCARDIV TWO (volume I)
COMCARDIV FOUR (volume I)
COMCARDIV SIX (volume I)
COMCARDIV FOURTEEN (volume I)
COMCARDIV SIXTEEN (volume I)
COMCARDIV TWENTY (volume I)

690-05-0011

UNCLASSIFIED

NOTE: FOR SPECIFIC CONTENT OF EACH VOLUME SEE INDEX OF ENCLOSURES
        WHICH IS ATTACHED TO INDIVIDUAL VOLUMES.

ALPHABETICAL LISTING OF WITNESSES AND PERSONS
WHO SUBMITTED STATEMENTS BEARING ON ALL MATTERS
LESS CASUALTIES AND INJURIES (See alphabetical
listing of fatalities for statements relating
thereto - Volume XII)

NAME	RANK RATE	VOLUME	ENCLOSURE
	AO3	VIII	(82A)
	CDR	VIII	(85)
	LT	IV	(30)
		VII	(52)
	GS-13	II	(13)
		VI	(51)
	LT	V	(42)
	LCDR	II	(6)
	AMHC	VII	(64)
	AN	IV	(38)
	LCDR	XI	(166)
	LI2	IV	(37)
	CAPT	IX	(89)
	LCDR	V	(44)
	LCDR	X	(112)
		XI	(172)
	LTJG	XI	(169)
	AMS3	VII	(62)
		X	(105)
	MMCS	V	(40)
	ABHAN	X	(103)
	LCDR	X	(91)
	LCDR	II	(3)
	AE3	X	(93)
	CDR	XI	(156)
	AO1	VIII	(81)
	LTJG	XI	(151)
	CDR	XI	(145)
	AA	VII	(66)
		X	(102)
	GMT1	XI	(180)
	MM1	XI	(186)
	AO3	VIII	(77)
	CWO2	VI	(47)
	PNSN	VII	(69)
	LTJG	X	(117)
	ABH2	XI	(185)
	CDR	IV	(28)
		VII	(53)
		VIII	(88)
	AMH1	III	(22)
	CDR	IV	(31)
		VIII	(86)
	AN	X	(98)
	W-1	XI	(182)
	AMHC	X	(119)
	AN	X	(99)
			(100)
	LCDR	II	(4)

All redactions are B.6.

NAME	RANK RATE	VOLUME	ENCLOSURE
	LCDR	X	(120)
	LCDR	IV	(36)
	LCDR (MC)	II	(14)
	AQC	XI	(170)
	AMS1	VIII	(62)
	AQ2	VIII	(80)
	SH3	VI	(46)
	AO3	III	(19)
		VIII	(75)
	AN	X	(107)
	AN	XIII	(403)
		II	(5)
	LCDR	X	(92)
	CDR	X	(133)
		XI	(157)
	AOAA	II	(16)
		VIII	(72)
	ABH1	XI	(168)
	AO1	VIII	(82)
	WO1	XI	(181)
	WO1	XI	(164)
	AQB2	XI	(150)
	MMC	XI	(184)
	AA	VII	(68)
	ADJ3	II	(15)
	AO3	X	(116)
	ENC	V	(39)
	RADM	X	(90)
	LT	XI	(178)
	AEC	IV	(33)
	CDR	V	(45)
		X	(134)
	AO3	V	(41)
		VIII	(79)
	AO1	III	(20)
		VIII	(74)
	LCDR	XI	(155)
	ABH3	XI	(154)
	AE3	X	(109)
	ABHC	X	(118)
	AN	XI	(149)
	LCDR	II	(8)
	AFCM	II	(9)
	LT	XI	(179)
	AA	X	(101)
	LTJG	II	(7)
	AO1	XI	(167)
	LCDR	II	(11)
	ABE3	IV	(32)
	LT	XI	(177)
	LCDR	VI	(48)
	AN	VII	(65)
	LT	XI	(162)
	ABHC	XI	(159)
	WO1	XI	(173)
	AEAN	X	(111)
	GS-11	VI	(50)
		VIII	(87)

All reductions are B.6

NAME	RANK RATE	VOLUME	ENCLOSURE
	AN	XI	(152)
	WO1	IV	(29)
	AO1	III	(21)
		VII	(59)
	AOC	XI	(147)
	LTJG	III	(25)
		VII	(57)
	CDR	VI	(49)
	ABCM	IV	(34)
	AA	VIII	(76)
	SF1	XI	(161)
	LTJG	X	(114)
	ENS	XI	(176)
	LTJG	II	(12)
	AQC	XI	(183)
	WO1	III	(24)
		VII	(56)
	CDR	X	(113)
	AO2	X	(115)
	ADJC	VII	(67)
	AN	XI	(153)
	LCDR	X	(126)
		X	(130)
	AN	X	(97)
	LT	X	(144)
	ABH3	X	(110)
	AT2	X	(104)
V.	AMC	X	(96)
	LT	XI	(171)
	ETN3	X	(106)
	AOC	III	(23)
		VII	(58)
	LCDR	XI	(165)
	LT	IV	(26)
		VII	(55)
		VII	(71)
	AN	XI	(163)
	CDR	II	(17)
	AOAN	XI	(148)
	AA	V	(43)
	LT	X	(108)
	AO2	III	(18)
		VII	(60)
		VIII	(73)
	WO1	X	(94)
	LTJG	X	(95)
	AO1	II	(10)
	MMC	XI	(174)
.	LCDR	IV	(27)
		VII	(54)
	AFCM	IV	(35)
	AO3	VIII	(82B)
	LCDR	XI	(146)
	AMHC	XI	(158)
	LTJG	XI	(175)
	AA	XI	(160)

All redactions are B-6

From: Senior Member, Informal Board of Investigation
To: Commander, Naval Air Force, U. S. Atlantic Fleet

Subj: USS FORRESTAL (CVA-59) Fire Investigation

Ref: (a) JAG Manual
(b) COMNAVAIRLANT MSG 092039Z August 1967

Encl: (1) COMNAVAIRLANT MSG 301617Z July 1967 appointing board as
modified by COMNAVAIRLANT MSG 031417Z August 1967
(2) Signed statement from each of the twenty parties acknowledging
advice of rights
(3) through (406) Testimony, statements and other evidence
(see index)

1. The informal investigation ordered by enclosure (1) to inquire into
the circumstances surrounding the FORRESTAL fire of 29 July 1967 has been
conducted in accordance with reference (a) and the report of said
investigation is hereby submitted.

PRELIMINARY STATEMENT

1. The members of the board, upon receipt of enclosure (1), proceeded
expeditiously from their respective stations at Quonset Point, R. I., Virginia
Beach, Virginia and Jacksonville, Florida to NAS Cubi Point, Republic of
the Philippines, all arriving on 3 August 1967. On that date the
Board was convened aboard USS FORRESTAL. Counsel for the Board, who
traveled from Atsugi, Japan, arrived at Cubi Point on 31 July 1967, as
FORRESTAL was standing in. Upon arrival, counsel for the Board met with
and commenced assisting the preliminary investigation that had been
ordered by COMCARDIV 2 immediately after the fire. That investigative
body, consisting of CARDIV 2 staff personnel, CAPT   B - 6
CDR   B - 6   CDR   B - 6   ENS   B - 6   and LT   B - 6
B - 6   FORRESTAL's Legal Officer, gathered evidence that proved
invaluable to this Board. CDR B - 6   , CDR B - 6   d ENS B - 6
continued to ably assist the Board throughout the investigation. The
decision to order an independent preliminary inquiry was prudent as it
not only produced important information which was immediately available
to the Board upon their arrival at Cubi Point, but also uncovered and
committed certain key witnesses in writing while events were still fresh

in their minds and before they had discussed the matters with others.

2. After a thorough reading and briefing of material developed by the
preliminary investigation, the Board was faced with the problem that
FORRESTAL would be departing Cubi Point for CONUS without means for
bringing additional personnel aboard once underway. This would require
the Board to rely entirely on the witnesses, counsel for parties and other
essential personnel who would be embarked for the 32 day transit. The
Board and counsel therefore were forced to devote almost the entire time
remaining in port to administrative matters and screening potential witnesses
so that statements and testimony could be taken from all who would not be
making the transit. The Board was greatly aided in this job by Assistant
Counsel, LCDR B-6   and LT . B-6   Flag Secretary and Aide
to the Senior Member. In addition LT B-6   ovided expert assistance
to LCDR B-6 in preparing the record of the investigation as did LT B-6
USNR, Legal Officer,   B-6   who arrived 11 August, and served as
an assistant counsel until 2 September providing much needed help during
that period.

3. Parties had to be designated as soon as possible and accorded their
rights in order that enough counsel could be obtained for those requesting
same before departure of the ship. The Board took testimony immediately
from those witnesses who were leaving the ship and by 6 August enough
information had been obtained to enable the Board to make considered
decisions as to who should be designated parties. Such determinations
were made on 6 August with the following twenty persons designated as
parties:

B-6 USN,	Commanding Officer, USS FORRESTAL (CVA-59)
✓	, USN,	Commanding Officer, VF-11
✓	, Jr., USN,	Commander, Carrier Air Wing 17

2

USN	Air Officer, USS FORRESTAL (CVA-59)
N	Engineering Officer, USS FORRESTAL (CVA-59)
USN	Carrier Air Wing 17 Senior LSO and pilot of aircraft #110 on 29 July 1967
USN,	Hangar Deck Officer, USS FORRESTAL (CVA-59)
USNR,	Damage Control Assistant, USS FORRESTAL (CVA-59)
Jr., USN	VF-11 Maintenance Officer
SN,	Carrier Air Wing 17 Ordnance Officer
USN,	Flight Deck Officer, USS FORRESTAL (CVA-59)
USN,	VF-11 Div. Off. Avionics/Weapons Div.
USNR	VF-11 Weapons Branch Officer
USN	Fire Marshall, USS FORRESTAL (CVA-59)
USN	Air Gunner, USS FORRESTAL (CVA-59)
SN	VF-11 Assistant Avionics/Weapons Div. Off.
SN,	VF-11 Weapons Branch Chief
USN	VF-11 Leader of an ordnance team
USN	VF-11 Plugged-in Rockets on aircraft #110
USN	VF-11 Conducted stray voltage test on aircraft #110

Since AO2   and AOAA   were patients at the Naval Hospital,
Subic Bay, R. I. counsel for the Board informed them of their designation
as parties at the hospital on the night of 6 August. Their rights were
also explained to them at that time and they both signed statements
acknowledging the explanation of their rights. (Enclosure (2)).

3

All redactions are B-6

Neither of these parties desired counsel at that time. The next morning, 7 August 1967, the remaining 18 parties were informed of their designation by the Board. Their rights were explained to them by counsel and they all signed statements acknowledging that their rights had been explained to them (Enclosure (2)). These 18 parties desired counsel certified in accordance with Article 27b, UCMJ. The Commanding Officer, USS FORRESTAL was the only party who requested counsel by name. He first requested CAPT   , USN and when CAPT   as declared to be not available a request was made for CDR   USN. When notification was received that he too was unavailable a request was made to the convening authority by CAPT   requesting a qualified and experienced law specialist. Ultimately LCDR   ', USN was made available and arrived aboard FORRESTAL on 10 August 1967. He served as counsel for CAPT   from that date throughout the investigation. Other counsel arrived during the period 7 August to 11 August and represented parties as follows, from the time they arrived throughout the investigation:

a. LT   SNR arrived 10 August 1967, from the Law Center, Norfolk, Virginia and served as counsel for CDR   LCDR 1 and LT .

b. LT   USNR arrived 10 August 1967 from COMFAIRWING THREE, NAS, Brunswick, Maine and served as counsel for CDR _____ , LT   and AO2

c. LT   USNR arrived 10 August 1967 from COMNAVPHIL and served as counsel for CDR   ?, LCDR   and LT

d. LT   , USNR arrived 10 August 1967 from NAS, Jacksonville, Florida and served as counsel for LTJG   AO2 and AOAA   ___ .

e. LTJG   , Jr. arrived 10 August 1967 from NAS, Jacksonville, Florida and served as counsel for CDR   , LCDF CWO2 .   nd WO1

4

All redactions are B-6

f. LTJG   rived 7 August 1967 from COMNAVBASE, Subic
and served as counsel for LCDR   , WO1   and AOC
All counsel were certified in accordance with Article 27b, UCMJ. On
10 August 1967 LT   on his own initiative talked to parties
and   at the Naval Hospital Subic and they changed their decision
in regard to representation by counsel. LT   represented them from
10 August throughout the investigation. All rights were fully accorded
all the parties. The Board in determining those who were to be designated
as parties looked at the duties and responsibilities that were inherent
in certain billets. If the duties and responsibilities appeared to have
a direct relationship to matters bearing on either the initiation of the
fire or the fighting of the fire, the individual filling the billet was
designated a party. If there was doubt in any case, the Board leaned
towards designation rather then against, to insure that the right to
counsel was afforded before getting underway.

3. Although considerable latitude in the manner of proceeding is afforded
an informal board, it was initially decided by the Board that a formal
hearing room procedure would be utilized whenever parties testified or
whenever it became apparent that a potential witness' testimony might
be controversial to any party. This was done to ensure that the parties'
rights were fully protected. In addition, witnesses who would not be
making the transit aboard ship were called to preserve their testimony
verbatim. The remainder of the testamentary evidence was gathered through
use of written statements. All statements were taken under oath when
practicable and if a witness, in addition, testified before the Board
he reaffirmed his statement, or statements, under oath. Several
witnesses made more than one written statement before they testified,
some of which were unsigned. All these statements were verified by the
witness when he testified and are included in this report in the enclosure
with the witness' testimony. All witnesses who testified before the

5

All redactions are B-6

Board did so under oath and each witness was informed of his rights under Article 31, UCMJ before he testified. In screening the witnesses to be called and the statements to be used in this report, the Board read approximately 1900 statements.

4. As soon as the Board reported to FORRESTAL a tour of the damaged areas was taken. After that, numerous places of interest on the ship were visited and revisited by the Board together and singly. The following are experiments which were conducted aboard ship at the Board's request and are described in the facts and enclosures:

    a. Two separate overall tests of flight deck fog foam stations.

    b. Static firing of a ZUNI rocket from an F-4B aircraft.

    c. Re-creation of what was reflected in the first few minutes of the Flat film by using flash bulbs on the after portion of the flight deck.

    d. Tests to determine the effectiveness of the LAU-10/A shorting device.

    e. Several tests of the electrical system of the F-4 and associated equipment by Mr. B b and CDR B b

5. One of the preliminary acts by the Board was to request two experts, one in ordnance and the other in the electrical field. In response to this request Mr. B b from Naval Weapons Center, China Lake, California and Mr. B b from Naval Missile Center, Pt. Mugu, California, reported aboard FORRESTAL at Cubi Pt., R. P. and stayed aboard until 2 September 1967. They greatly assisted the Board throughout the investigation and their testimony and statements appear as enclosures (50), (51) and (87). These enclosures must be read for a full understanding of the events of 29 July 1967 aboard FORRESTAL. As an additional aid to the reader there is attached to the preliminary statement a glossary of terms, colloquialisms, and abbreviations used throughout this report and a compendium on the circuits and equipments which are pertinent to the firing of a ZUNI rocket from a LAU-10/A launcher suspended from a TER-7

6

on an F-4B aircraft. Both of these documents were prepared by CDR. ß-6
ß-6 USN, whose service as a technical assistant and general advisor
to the Board was immeasurable.

6. There were certain conflicts that arose in the testimony during the
hearings and the Board resolved these conflicts based on all the evidence,
together with evaluations of the witnesses as they appeared and reacted
to questioning. All the conflicts that were considered important by the
Board have been clearly resolved in the findings of fact and opinions.
Testimony that was considered to be false is so labeled in the opinions.
No disciplinary action is recommended for those witnesses who are believed
to have testified falsely, however, since it is considered that proof of
an offense would be almost impossible.

7. One apparent conflict, or question, is left dangling in the testimony
without an answer. That is the part played by ß-6 , AO1, in the
sequence of events on 29 July. ß-6 , a member of Fighter Squadron
11, was the safety petty officer in charge of the composite catapult
arming crews while FORRESTAL was on the line.

It was thought at one point during
the investigation that ß-6 might have some important information to
offer to the Board and several questions were put to certain witnesses
concerning ß-6 Pursuant to dispatch request by the Senior Member
information was ultimately received in the mail while at sea summarizing
an interview of ß-6 conducted at Portsmouth Naval Hospital by an
agent of ONI. This report revealed that ß-6 's testimony, if called
as a witness, would not differ in marked degree from any other VF-11
witness who was called.

8. By reference (b) authority was granted to omit the requirement for
line of duty/misconduct determinations for all those injured by the fire

7

and explosions. In accordance with reference (b) injury reports and medical record entries are to be made by Commanding Officers of injured personnel and appropriate medical commands. Although detailed medical documentation of all injuries has not been included in this report, a summary of all injuries can be found in enclosures (328) and (329) and an LOD/Misconduct opinion is expressed in regard to the injuries.

9. One of the most important items of evidence utilized by the Board was the Flat tape from the morning of 29 July 1967. At the time the fire started the Flat camera was trained on a ~~KA-B~~ *KA-3B* being readied for launch from the port bow catapult. While trained in this direction the camera photographed the flame and smoke from the ZUNI rocket which was reflected from the plexiglass Flat booth window. The Flat camera was then turned aft and trained on the burning A-4's on the port quarter. The camera remained in this position throughout the fire on the flight deck and the resultant tape presents vivid evidence of the fire and explosions. The original Flat tape and a kinescope copy of this tape are forwarded with the original report of investigation.

10. The preliminary statement is supposed to reveal the difficulties encountered in the investigation. No difficulties mentioned can overshadow the sheer weight of clerical problems involved in producing a thirteen volume record. Much assistance was furnished by FORRESTAL but the real credit for producing this record goes to:

YNC		COMFAIRQUONSET
YNC	B-b	USS FORRESTAL (CVA-59)
YNC		NAS KEY WEST
YN1	HS-2	

Their talent at reporting and transcribing, along with all their other myriad yeoman abilities made production of this record a matter that was never a real problem for the Board. Their competent assistance was of the highest value to the Board.

GLOSSARY OF TERMS, COLLOQUIALISMS AND ABBREVIATIONS

Numerous terms used in testimony and introduced into the record were incomplete, colloquial in nature, or inaccurate. In view of the technical nature of many of the equipments or functions which were fundamental to the investigation, it was considered advisable to prepare this section to provide the reader with a clear understanding of terms of reference.

Definitions or descriptions, synonyms or colloquialisms of the most important terms associated with carrier operation of the F-4B aircraft/Triple Ejector Rack/LAU-10/A Launcher combination have been included. The descriptions of ordnance items are related specifically to the production model actually in use at the time of the accident.

Several official publications in which a term appeared did not always use consistent nomenclature. Definitions for some terms were not provided. The board therefore made arbitrary selections of definitions in those cases. Some abbreviations commonly used aboard carriers have also been included and will be used in the body of the report without further amplification.

Aero 3A Launcher	A rail launcher with detent and contact point for carrying and launching an AIM - 9 B (Sidewinder) missile. Usually mounted on the lower outboard side of the LAU-17/A pylon of the F-4B aircraft but may also be mounted on the inboard side. When installed, a safety pin interrupts the electrical firing circuit. Also referred to as Aero 3, S/W launcher.
Aero 7A Launcher	A launcher which carries an AIM-7 (Sparrow III) missile in a semi-submerged position at one of four stations under the fuselage of an F-4B aircraft. Incorporates a launcher safety pin which prevents inadvertent ejection of the AIM - 7 missile.
Aero 27A Bomb Rack	The centerline (external stores station 5) ejector bomb rack of the F-4B aircraft. Carries a variety of external stores including a 600 gallon fuel tank or a MER or TER.
AN-M65 Bomb	General purpose 1000 lb bomb of World War II design, properly designated AN-M65A1. Bombs of this type aboard FORRESTAL were loaded with 555 lbs of Composition B in 1953. Also referred to as "fat bomb" because of high aerodynamic drag characteristics.
Armament Safety Override Switch	Switch with holding relay supplied from right 28 V DC bus. Mounted above left rear console in front cockpit. Bypasses Landing Gear Control Armament Safety. Used for circuit continuity checks. Also called armament override switch.

9

"Bomb Farm"	A colloquialism used in reference to a designated area on either the flight or hangar deck where unfuzed bombs are collected temporarily until either struck below into magazines or loaded on aircraft.
Bomb Release Switch	Final switch in bomb/rocket release circuit. Located on upper left portion of pilot's control stick grip. Commonly referred to as pickle, bomb pickle, bomb button, firing button.
"Cat (s)"	Catapult(s).
CBU Rocket Switch	A three-positioned switch located in the tail cone of a TER (or MER) which controls the mode of release or firing of CBU and rockets.
Composition B	The explosive mixture contained in AN-M65A1 bombs and the MK24 warheads of the ZUNI. The mixture ratio is 59% RDX, 40% TNT and 1% wax.
"Conflag " Station	Conflagration station. A compartment in a hangar bay located well above deck level to give an unobstructed view of the deck below. Equipped with controls to close hangar bay doors and activate fire-fighting systems.
"Dog Bone"	See Weapons Control Panel
Ejector Rack Safety Pin	Pin providing a mechanical safety for each ejector rack hook. If pin can be inserted after loading, locking of hooks is assured. Each ejector rack requires two pins. Also referred to as rack pin, mechanical pin, mechanical safety pin, L-shaped pin.
Explosive Bolt	See LAU-17/A Explosive Bolt
Frangible Fairing	Composition paper cover attached to the forward and aft ends of rocket launchers to reduce aerodynamic drag. The fairing is destroyed readily when the first rocket in the launcher is fired. Also referred to as fairing, cover, nose dome, dome, nose cover.
H-6	An explosive mixture contained in MK82 and M117 general purpose bombs. The mixture ratio is 40% RDX, 38% TNT, 17% Aluminum and 5% wax.

-2-

HCFF	High Capacity Fog Foam System designed for combating severe "Class B" fires on the hangar deck and flight deck. Intended to provide large quantities of foam in a relatively short time. There are 17 HCFF foam generators, with 300 gallon foam liquid reservoirs, installed on the second deck. These large generators are designed to furnish fog foam to monitors and hoses on the hangar deck and to hoses on the flight deck. Individual stations are activated remotely from hangar and flight deck stations by opening a stop valve and energizing the system by depressing an electrical switch.
Home-Step Switch	This three-position toggle switch is mounted on the tail cone at the rear of the TER-7 or MER-7. The circuit is energized whenever aircraft power is available. Placing the switch in the HOME position will cause the internal stepper to rest on the number one station of the rack and the homing light to illuminate. If the switch is then placed to the OFF position, the rack will deliver sequential single or automatic firing or release impulses according to the setting on the Release Mode Selector Switch. If the switch is placed in the STEP position, the next station in sequence will be selected, facilitating circuit continuity ground checks. Although the functions remain unchanged, the switch positions for HOME and STEP in the -527 TER/MER are reversed from those in the -505 or -521 TER/MER.
"Huffer"	Colloquial term for an aircraft starting tractor.
HYTROL (Valve)	Acronym for hydraulic control. Used alone to refer to a valve in the HCFF system.
In the pack	The location of an aircraft on the flight deck, usually in close proximity to other aircraft, where there is no assurance that the area ahead of the aircraft will not be obstructed by people or other aircraft between the time the aircraft is manned and the time it is taxied forward for launch.
Intervalometer	An electrical device for timing the firing of rockets or release of bombs. Individual intervalometers are incorporated in the F-4B Weapons Control System, the TER, and the LAU-10/A launcher.
Intervalometer pin	See TER Electrical Safety Pin.

JBD	Jet Blast Deflector. A two-positioned steel ramp which is raised behind aircraft being launched from a catapult to prevent high velocity jet exhaust gas from travelling down the flight deck.
LAU-3A/A Launcher	A 19 round launcher for 2.75 in. rockets. An rf filter, an rf barrier and a breaker switch make it safe for use in a RADHAZ environment. The breaker switch function is similar to that of the shorting device in the LAU-10/A. Also called LAU-3, LAU-3/A.
LAU-7/A Missile Launcher	Launcher for the AIM-9B or -9D Sidewinder missile. Mounted on the lower inboard or outboard side of the LAU-17/A pylon. Also called the LAU-7 launcher.
LAU-10/A Launcher	A four round launcher for the 5.0 in ZUNI rocket. Two electrical receptacles in line with removable suspension lugs except either a shorting device or a rocket harness. A selector switch mounted on the rear bulkhead provides a choice of single or ripple fire. An intervalometer mounted inside the forward bulkhead moves off a normally shorted position if more than 5 volts DC and 2 amperes are introduced into the firing circuit and causes one or all four rockets to fire depending on the position of the selector switch. After the launcher is suspended from the ejector rack, frangible fairings may be attached at both ends of the launcher to reduce aerodynamic drag. Also called ZUNI pod, ZUNI Launcher, ZUNI LAU-10 pod, LAU-10, ZUNI, rocket pod.
LAU-10/A Receptacle	Dual purpose receptacle which accepts either a shorting device or a rocket harness. Each LAU-10/A launcher has two receptacles, one forward and one aft. Also called receptacle, rocket receptacle.
LAU-10/A Selector Switch	A two-position toggle switch mounted on the aft bulkhead of the LAU-10/A launcher to provide for alternate selection of single or ripple fire.
LAU-17/A Pylon	Wing pylon attached to inboard under side of wing by an explosive bolt forward and a support attachment aft. Contains the male portion of the pylon electrical disconnect in the upper forward portion of the pylon. Carries one AIM-7 missile on rails or one or two AIM-9 missiles on Aero 3A or LAU-7/A launchers mounted on the lower inboard and outboard sides of the LAU-17/A. When a pylon adapter is bolted on to the LAU-17/A pylon, which precludes carriage of an

—

12

AIM-7, either a TER or a MER can also be carried. When installed, the LAU-17/A Safety Pin prevents current from reaching the explosive bolt and, if carried, the AIM-7 motor, but does not affect electrical circuits of the TER or AIM-9 missiles. Also called Inboard Pylon, Wing Missile Pylon, Pylon, Wing Pylon, LAU-17.

LAU-17/A Pylon Adapter	Provides a means of suspending a TER or MER from the LAU-17/A pylon.
LAU-17/A Pylon Explosive Bolt	An explosive bolt in the pylon which, when fired, causes the pylon to separate from the wing. Also called Explosive Bolt.
M117 Bomb	A 750 lb. general purpose bomb of post-World War II design, properly designated M117A1. Contains 393 lbs of H-6 explosive. Also referred to as 117.
MER	A/A 37B-6 Multiple Ejector Rack. Part numbers of the three Type 7 models are 5821500-505, -521, -527. Consists of six bomb ejector units arming solenoids, control units and circuits which enable it to carry, arm and eject six MK81 or MK82 series bombs, or carry, fire and eject three rocket launchers, or carry, dispense and eject three CBU. Also called MER, MER-7, MER 527 or 527 MER.
Mezzanine	Term used to refer to the 01 level aboard carrier, particularly those areas or compartments aft with direct access to the hangar deck.
Mickey Mouse	Colloquial term used by witnesses in reference to sound attenuators or to entire flight deck helmet which includes sound attenuators. Synonomous with colloquial term "Mickey Mouse ears" or "ears".
Missile	The word used commonly to refer to a guided, rocket propelled missile such as Sidewinder, Sparrow or Shrike.
MK 82 Bomb	A slender 500 lb bomb of post-Korean War design with low aerodynamic drag characteristics. Loaded with 192 lbs of H-6 explosive.
Non-propulsive Attachment	A device which is locked into the exit cone of a rocket nozzle in such a way as to cause the thrust to be neutralized. Used only on the AIM-9B Sidewinder missile. Also called NPA, Non-propulsive unit, NPU, thrust neutralizer.

OBA | Oxygen Breathing Apparatus. Provides the wearer with a supply of oxygen and respiratory system independent of the surrounding atmosphere. Oxygen is provided by chemical contained in a replaceable metal cannister. Naval personnel engaged in fire-fighting replace the cannister every 30 minutes when the timer on the apparatus sounds a warning signal.

"Pickle" | See Bomb Release Switch.

"Pigtail" | See Rocket Harness

Pylon Electrical Disconnect | The mating connections used to complete all electrical circuits between the aircraft wing and the LAU-17/A pylon. Consists of a receptacle (part number 32-75031-3) in the wing, a plug (part number 32-75031-5) in the pylon and an adapter (part number 32-75031-7) which provide a clean jettison capability and facilitate installation. Also referred to as the 101 pin connector, 101 connector.

Rocket | The word used commonly to refer to an unguided missile propelled by the reaction from gases formed by burning propellant.

Rocket Harness | Cable used to make an electrical connedtion between the TER and the LAU-10/A launcher. Commonly called the pigtail, rocket cable, rocket pigtail.

Safe-Arm Switch | See shorting device. Also may refer to a lever on the AIM-7 missile used to interrupt the motor firing voltage.

Shorting Device | A sliding switch on the LAU-10/A Launcher which in the SAFE Position causes electrical signals to be shorted to ground. The sliding action of the switch handle causes, through a ramp and spring, the insertion or removal of a metal cone between the pins in the receptacle. Locks in the armed position but does not have a detent in the safe position. Shorting devices are installed in both LAU-10/A receptacles during production. Loading or assembly crews remove device from the receptacle which is to be used for connection with the rocket harness. Also called LAU-10 safety, slide switch, safety and arming device, safe-arm switch, shorting switch.

Sidewinder	Sidewinder is a guided missile which homes passively on infra-red energy. Two models designated AIM-9B and AIM-9D are carried by F-4B aircraft. The LAU-7/A launcher is compatible with both models but the Aero 3A launcher can only carry the AIM-9B. Also referred to as AIM-9, S/W, S/W-9B, or S/W-9D.
Sparrow	Sparrow is a semi-active radar homing missile designated AIM-7E. Although Sparrows can be launched from a LAU-17/A pylon of the F-4B aircraft, the missiles are normally carried on fuselage Aero 7A launchers in order to facilitate the carriage of Sidewinders and TERS on the LAU-17/A pylon. Also referred to as AIM-7, SP III, SP 7E.
"SPUD"	Spray nozzle attached to the end of a salt water applicator.
Stray Voltage	An unwanted voltage which has been imposed on the firing circuit of a weapon system. If not detected, arming of the weapon system allows the voltage to be passed to the initiating element of the weapon. Depending on its magnitude inadvertent firing of the weapon may result.
Tanker	An aircraft capable of refueling another aircraft in flight. Used primarily to mean a KA-3B aircraft.
TER	A/A37B-5 Triple Ejector Rack. Part numbers of the three Type 7 models are 5821520-505, 5821520-521, and 5821520-527. Consists of three bomb ejector units, arming solenoids, control units and circuits which provide the flexibility required to carry, fire, and eject rocket launchers, carry, arm and eject bombs, and carry, dispense and eject CBU. Also called TER, TER-7, and referred to as 527 TER, TER 527 or new TER.
TER Electrical Safety Pin	The pin (part number 4815967-1) which is installed in a receptacle in the tail cone of either a TER or MER to interrupt the firing or release circuit. Also referred to as TER pin, TER safety pin, Intervalometer pin.
TER Safety Pin	See TER Electrical Safety Pin
Transient Voltage	A momentary increase (or decrease) in normal circuit voltage caused by interrupting the circuit, changing the source of the current, or changing the load on the circuit.

Umbilical	An electrical cable or wafer used to connect a missile to a launcher.
Weapons Control Panel	Panel in the F-4 front cockpit used to select and control the release or firing of weapons carried on the MERS or TERS. Also referred to as the "dog bone".
Weapons Switch	Two position toggle switch on Multiple Weapons Control Panel which performs the functions of a master armament switch. Interrupts circuit between the Bomb Release Switch and all MER/TER when in CONV OFF-NUCL ON position because the bomb transfer relay is not energized. Referred to as Master Arm, Master Arm Switch, Master Armament Switch and Armament Master Switch.
Yankee Station	Area in the Gulf of Tonkin used by TF 77 attack carriers when launching offensive strikes against targets in North Viet Nam.
Yellow Equipment	Generic term for equipments used in support of ground test and operation of aircraft. Such as tractors, starters, mobile avionics or hydraulic test equipments, oxygen carts, bomb skids, wheel chocks, tie-down chains, tow bars and boarding ladders. Also referred to as yellow gear.
ZUNI	A five inch folding fin aircraft rocket with a variety of warheads and fuzes which is carried in a LAU-10/A launcher. Commonly used as a flak suppression weapon by installing an M414A1 VT fuze in a MK 24 General Purpose Warhead.
ZUNI rod, ZUNI Rocket Pod	See LAU-10A Launcher

F-4B Aircraft - Zuni Rocket

A study of the publications and documents listed below was conducted by CDR | _ B-6   USN, 498079/1310, Air Warfare Officer, Staff, Commander Carrier Division TWO, to produce this compendium on the equipments and circuits which are pertinent to the firing of a Zuni rocket from a LAU-10/A launcher suspended from a TER-7 at external stores station number 2 of a 153 --- series F-4B aircraft.

Bibliography

"Aircraft Rockets" NAVWEPS OPORD 2210 Volume 1 (First Revision of 15 September 1966).

"Conventional Weapons Loading Manual - Navy Model F-4B Aircraft", NAVAIR 01-245FDB-75 (Interim Manual Change No. 6 of 9 September 1966 entered).

"Maintenance Instructions Manual - Navy Model F-4B, F-4J and RF-4B Armament Systems", NAVAIR 01-245FDB-2-7 (1 July 1966).

"NATOPS Flight Manual Navy Model F-4B Aircraft", NAVAIR 01-245FDB-1 (Changed 1 May 1967).

"Operation, Maintenance and Overhand Instructions, Triple Ejector Rack (Type 7)", NAVAIR 11-75A-40 (Changed 15 Apr 67).

Statements and Testimony of Mr.   B-b   (Contained in Vol I and Vol V of the Record).

Statements and Testimony of Mr. B-b   (Contained in Vol V and Vol VII of the Record).

External Stores Station Number Two of F-4B Aircraft #110

A brief description of the mechanical and electrical design features of the equipment which was installed at external stores station #2 on the port wing of F-4B aircraft #110 will aid in understanding how a Zuni rocket could have been fired from that station.

ENCLOSURE ( )

17

LAU-17/A Inboard Pylon. The LAU-17/A pylon, the basic building block of the inboard station, was designed to carry and launch one AIM-7 (Sparrow) missile from its rail or, alternatively, one or two AIM-9 (Sidewinder) missiles from either Aero 3A or LAU-7/A launchers mounted on the sides of the lower portion of the pylon. The LAU-17/A pylon is attached to the wing by an explosive bolt forward and a mounting cam aft. When its jettison circuit is energized, either individually by using the missile control panel or along with all other stations by activation of the external stores emergency release switch, the pylon and any missiles or other stores attached separate from the wing. To aid clean separation, all of the electrical circuits entering the pylon are collected in the pylon electrical disconnect, a 101 pin cannon plug with mating adapter. The LAU-17/A has a receptacle for a safety pin which, when installed, interrupts both the jettison circuit and the Sparrow motor firing circuit. Another receptacle on the side of the launcher is used to check for stray voltage in those two circuits before the safety pin is extracted.

LAU-7/A Launchers. The LAU-7/A is compatible with both the AIM-9B and AIM-9D missiles. The LAU-7/A has a safety pin which, when installed, locks the AIM-9 in place by mechanically preventing movement of the missile detent and also prevents completion of the firing circuit by mechanically holding two solenoids in an open position.

-527 TER When other than AIM-7 or AIM-9 missiles are to be carried on the inboard wing stations, a TER is suspended from an adapter which is bolted to the LAU-17/A pylon. Like other triple ejector racks, the -527 TER consists of three 14 inch suspension ejector racks with Aero 2B arming units, a control unit, and wiring circuits required to fire rockets, despense CBU bomblets and eject bombs, rocket launchers or CBU containers. Bombs may be ejected in either an armed or safe condition. The tail cone assembly at the aft end of a wire bundle support assembly houses a home-and-step control toggle switch, a press-to-test homing indicator light, a release mode selector switch, and a receptacle for an electrical safety pin. Harness assemblies which pass through the adapter to connect with terminals in the LAU-17/A pylon supply power to homing and stepping circuits of the -527 TER whenever aircraft power is available. If the stepper circuit is homed, power is directed first to TER station #1 (center line) and then sequentially to station #2 (left shoulder looking forward), and #3 (right shoulder). Power is supplied to each LAU-10/A rocket launcher by a rocket harness (commonly called "pigtail") which connects the appropriate receptacles at the after portion of the TER to the aft receptacle of each launcher. The LAU-10/A launcher can be ejected from the TER, in which case the rocket harness will be retained by a wire bale connected to the TER receptacle.

LAU-10/A Launcher. This launcher is used as a shipping container for four Zuni rocket motors and as a launcher for four complete Zuni rounds. In addition to five suspension lug wells to make it compatible with a variety of bomb racks, the launcher has a spring-loaded detent mechanism for each tube, an intervalometer, a selector switch for pre-flight selection of either single or ripple fire, and two five-pin electrical receptacles wired in parallel which can accept either a shorting device or a rocket harness. When the LAU-10/A launcher is to be suspended from a TER, the aft shorting device is removed to permit connection between the rocket harness and the launcher. Since the receptacles are wired in parallel, any electrical power which passes through the rocket harness should be shunted to ground if the remaining shorting device is in the safe position. Any electrical power

690105-0011

which passes the shorting device will be directed to the intervalometer which, in the zero position, grounds all of the rocket squibs. If the power exceeds approximately 10 watts, the intervalometer will move from the grounded position to station #1 and direct power to the contact band of the rocket in launcher tube #1 (lower left, looking forward). Since only approximately .08 watts are required to activate the motor ignition squib, a Zuni rocket will be fired whenever the impulse to the launcher has been sufficient to cause the intervalometer to step. If the selector switch has been set on ripple and the firing pulse lasts approximately one-half second, all four rockets will be fire by a single pulse.

F-4B Aircraft Rocket Firing Circuit

Assuming that no malfunctions exist, there are nine conditions which must be fulfilled in order to complete the electrical circuit to fire a Zuni rocket from an F-4B aircraft on deck. Although the conditions are listed below in a logical order, the sequence is not important from an electrical standpoint, since the circuit is in series. The conditions, most of which are actually specific switch positions, are:

1. Electrical power available to the aircraft distribution system.

2. Weapons Switch in CONV-ON-NUCL-OFF position.

3. Weapons Selector Switch in RKTS DISP position.

4. Station Selector Switch in INBD WING position.

5. Rocket harness connected to LAU-10/A launcher.

6. TER electrical safety pin removed.

7. Shorting device in the armed position.

8. Armament Safety Override Switch depressed.

9. Bomb Release Switch depressed.

Aircraft Power. Electrical power may be supplied from either an external source or from either or both of the engine driven alternating

-4-   ENCLOSURE (   )

20

current (AC) generators which supply three phase 115/200 volt 400 cycle AC power and, using transformers and rectifiers, 28 volt AC and direct current (DC) power to the aircraft distribution system. Excepting possibly during the brief period of transition from one power source to the other when the numerous magnetic fields of electrical motors, generators, and wiring throughout the aircraft will be collapased and reconstituted resulting in inductive currents, electrical power is distributed in the same manner irrespective of the original source.

Multiple Weapons System Control Panel. This panel provides control for firing, release, sequencing, and arming of the various external stores carried on stations 1, 2, 5, 8 and 9. The panel, mounted in front of the pilot's control stick, contains a Weapons Selector Switch, Station Selector Switch, Weapons Switch, Step Reset Switch, Bomb Arming Switch, and Interval Selector Switch. Only the Weapons Selector Switch, Station Selector Switch and Weapons Switch affect the ability to fire a rocket.

Weapons Switch. Power is not supplied to either the CONV-ON-NUCL-OFF or CONV-OFF-NUCL-ON position of this switch which performs the function of a master armament switch, unless the landing gear handle is raised or the Armament Safety Override Switch is depressed. These two provisions are incorporated for the specific purpose of preventing inadvertent release of external stores or firing of rockets while retaining the capability of performing circuit continuity checks when desired.

Weapons Selector Switch. The position of this rotary switch controls the type and release method of external stores. This switch must be in the RKTS DISP position in order to permit normal firing of a rocket. This switch is customarily placed in the BULL PUP position when not in use since that circuit is inactivated when the Aero 5A launcher for the BULL PUP missile is not installed. Placing this switch in the RKTS DISP

-5-

position permits power to close the select rockets relay incorporated in each TER.

Station Selector Switch. This rotary switch has five positions in use: OFF, OUTBD WING (#1 and #9), INBD WING(#2 and #8), CTR (#5), and all (#1, #2, #5, #8 and #9). Reject relays are energized when this switch is in the OFF position and prevent normal station selection signals from going to any of the stations.

Bomb Release Switch. This switch, commonly called the "pickle", is a spring loaded button mounted on the left upper portion of the control stick grip. When depressed, with power available from the armament bus, the circuit from the armament bus to the bomb switch transfer relay will be closed and power will be directed to either the special weapons release circuit or the multiple weapons release circuit. The bomb release switch is used to fire rockets as well as to release bombs depending on the position of the Weapons Selector Switch

Armament Safety Override Switch. This spring-loaded push type switch, mounted above the left console at the rear of the pilot's cockpit, is used to bypass the armament safety feature of the landing gear control when it is desired to make continuity checks of the bomb release and rocket firing circuits. Once depressed with power available, the switch is held in place until its holding coil is de-energized by the interruption of electrical power.

LAU-17/A Inboard Pylon. This pylon has been described previously.

-527 TER. This rack has been described previously.

LAU-10/A Launcher. This launcher has been described previously.

Electrical Safety Features Associated with the Armament Systems

Armament Safety feature in the landing Gear Control. A cam-operated switch in the landing gear control mechanism serves to interrupt all

-6-

power to the armament bus when the control handle is in the down position. The position of this switch does not affect the capability to jettison external stores or fuselage missiles. It should also be noted that the homing and stepping circuits of the TER are energized whenever aircraft power is available.

Left Main Gear Scissor Switch. The position of this sealed unit plunger switch mounted on the upper scissor link of the left main landing gear, controls the capability to jettison all fuselage missiles and all external stores. When the landing gear strut is fully extended, the switch closes completing a circuit from the 28 volt DC Essential Bus to the EXT STORES EMER REL switch mounted on the pilot's forward left console. The position of this switch has no effect on normal release or firing circuits which are completed by closing the Bomb Release Switch.

Emergency Stores Switch. This switch directs 28 volt DC power from the Essential Bus to energize the EXT STORES EMER REL Switch when the landing gear control handle is raised. This switch does not affect normal release or firing circuits.

FINDINGS OF FACT

SECTION I

BACKGROUND AND ENVIRONMENT

1. That on the morning of 29 July 1967, USS FORRESTAL (CVA-59), Captain ⟨B-6⟩ ISN, 105702/1310, Commanding Officer, was operating on Yankee Station in the Gulf of Tonkin as a part of Task Group 77.6.

2. That FORRESTAL chopped to COMSEVENTHFLT at 080001Z July 1967.

3. That FORRESTAL arrived on Yankee Station 24 July 1967, and first launched strikes against targets in North Vietnam at 0600H, 25 July 1967.

4. That on the morning of 29 July, Commander Carrier Division TWO, Rear Admiral ⟨B-6⟩ ISN, 078676/1310, was embarked with his staff as CTG 77.6, Commander Attack Carrier Striking Group, comprised of USS FORRESTAL (CVA-59), USS RUPERTUS (DD-851), with Commander Destroyer Division THIRTY-TWO embarked, and USS TUCKER (DD-875).

5. That friendly U. S. Naval carrier forces in the near vicinity were:

TASK DESIGNATION/COMMANDER	FLAGSHIP	ESCORTS
a. CTG 77.0/CTG 77.8 COMCARDIV NINE RADM ⟨B-6⟩	USS ORISKANY (CVA-34)	USS MACKENZIE (DD-836) USS MOORE (DD-747)
b. CTG 77.7 COMCARDIV SEVEN RADM ⟨B-6⟩	USS BON HOMME RICHARD (CVA-31)	USS FECHTELER (DD-870) USS SPROSTON (DD-577)

6. That CTG 77.5, Captain ⟨B-6⟩ embarked in USS INTREPID (CVS-11) with USS BAUSELL (DD-845) and USS PORTERFIELD (DD-682) in company was enroute Yankee Station.

24

7. That CTF 77, Commander Attack Carrier Striking Force SEVENTH Fleet, RADM 〈R-b〉 embarked in USS CONSTELLATION (CVA-64), was moored at Subic Bay, R. P.

8. That Commanding Officer, USS FORRESTAL (CVA-59), was Commander Task Unit 77.6.1, and was OTC (Officer in Tactical Command) of that unit.

9. That USS FORRESTAL (CVA-59) was engaged in high tempo combat operations involving air strikes against targets in North Vietnam.

10. That Attack Carrier Air Wing SEVENTEEN (CVW-17), comprised of the following squadrons and a detachment, was embarked in USS FORRESTAL (CVA-59):

    a. Commander Attack Carrier Air Wing SEVENTEEN and staff.

    b. Fighter Squadron 11 (VF-11).

    c. Fighter Squadron 74 (VF-74).

    d. Attack Squadron 46 (VA-46).

    e. Attack Squadron 106 (VA-106).

    f. Attack Squadron 65 (VA-65).

    g. Reconnaissance Heavy Attack Squadron 11 (RVAH-11).

    h. Carrier Airborne Early Warning Squadron 123 (VAW-123).

    i. Heavy Attack Squadron 10 Detachment 59 (VAH-10 DET 59).

11. That detachments embarked in and attached to the ship were:

    a. Helicopter Combat Support Squadron 2 Detachment 59 (HC-2 DET 59)

    b. Helicopter Antisubmarine Squadron 2 Detachment 59 (HS-2 DET 59)

    c. Carrier Airborne Early Warning Squadron 13 Detachment 59 (VAW-13 DET 59)

    d. Fleet Air Reconnaissance Squadron 1 Detachment 59 (VQ-1 DET 59)

12. That the aircraft assigned to the squadrons and detachments embarked in USS FORRESTAL (CVA-59) and the numbers actually on board, airborne or based ashore at 1051H, 29 July 1967, were:

Squadron or Detachment	Type Aircraft	Number Assigned	Number on Board	Number Airborne	Number Ashore
VF-11	F-4B	12	12	0	0
VF-74	F-4B	12	10	2	0
VA-106	A-4E	14	14	0	0
VA-46	A-4E	14	14	0	0
VA-65	A-6A	9	8	0	1
RVAH-11	RA-5C	6	5	0	1
VAW-123	E-2A	4	2	0	2
VAH-10 DET 59	KA-3B	5	2	2	1
VQ-1 DET 59	EA-3B	1	0	1	0
VAW-13 DET 59	EA-1F	2	1	1	0
HC-2 DET 59	UH-2A	3	3	0	0
HS-2 DET 59	SH-3A	4	2	2	0
CVA-59	C-1A	1	0	1	0

13. That weather and sea conditions were:

    a. Cloud cover 2000 feet scattered, higher broken.

    b. Visibility 10 miles.

    c. Temperature 87 degrees F.

    d. Dew point 78 degrees F.

    e. Humidity 74%.

    f. Wind 030 degrees true, 6 knots.

    g. Sea state calm.

    h. Sea temperature 88 degrees F.

    i. Altimeter setting 29.60.

14. That FORRESTAL was in material condition Yoke which had been checked at 0700H, 29 July 1967.

15. That on 29 July 1967, the fifth day of combat operations on Yankee Station, FORRESTAL was scheduled to launch strikes as shown in the Air Plan, enclosure (124).

16. That aircraft on deck for the 1100H scheduled launch were actually loaded (see enclosure (13) to enclosure (89)) with ordnance as follows:

NO. & TYPE A/C	SIDE NOS.	ORDNANCE LOADING
8 A-4	306, 316, 405, 407, 410, 414, 416, 417	2 AN-M65A1 1000 lb. bombs 35 rounds 20mm ammunition
3 A-4	301, 310, 314	2 MK82 500 lb. bombs 2 AGM-45 (Shrike) missiles 35 rounds 20mm ammunition
1 A-4	412	4 M117 750 lb. bombs 35 rounds 20mm ammunition
3 F-4	110, 202, 212	2 AIM-9B (Sidewinder) missiles 2 AIM-7E (Sparrow III) missiles 6 LAU-10/A launchers with 4 ZUNI rockets each
2 F-4	106, 204	One AIM-9D missile 2 AIM-7E missiles 6 LAU-10/A launchers with 4 ZUNI rockets each
2 F-4	101, 112	One AIM-9B missile One AIM-9D missile 2 AIM-7E missiles
2 F-4	105, 206	2 AIM-9D missiles 2 AIM-7E missiles
1 F-4	200	One AIM-9B missile 2 AIM-9D missiles 2 AIM-7E missiles
1 F-4	210	2 AIM-9B missiles One AIM-7E missile
1 F-4	113	2 AIM-9B missiles 2 AIM-7E missiles
3 A-6	505, 510, 511	18 MK82 500 lb. bombs 2 LAU-10/A launchers with 4 ZUNI rockets each

17. That because of the types of ordnance exposed to radiation and the types of radiation present, RADHAZ measures were not required and no RADHAZ condition was in effect.

18. That flight quarters on 29 July 1967 was sounded at 0400H.

19. That at 1025H pilots were directed to man aircraft.

20. That the aircraft spot at 1051H was approximately as shown in enclosure (10) to enclosure (89).

21. That at 1046H FORRESTAL turned into the wind, preparatory to launch, coming to course 050 degrees T, speed 27 knots.

22. That on launching course, wind over the deck was 32 knots from 350 degrees relative.

23. That at 1051H crews had manned aircraft and starting was in progress; some were started, some were not. Two aircraft, one KA-3B and one EA-1F, had been launched at 1050H. Launch of a second KA-3B and one E-2A was in progress. Above launch was preliminary to the 1100H launch.

24. That at 1051H A-4E #405, pilot LCDR Fred D. WHITE, USN, 627870/1310, was in the third aircraft forward of the stern on the port side of the flight deck, with engine running.

25. That A-4 #405 carried the following stores:
    a. 400 gal. centerline fuel tank. (full)
    b. 2 AN-M65A1 1000 lb. bombs.
    c. 35 rounds 20mm gun ammunition.

26. That at 1051H men were positioned around F-4B BUNO 153061 modex #110 approximately as shown in enclosure (139), starting the aircraft, checking it, and preparing it for flight. Their functions at this time were:

Name	Assigned Duty	Duty being performed at 1051H, 29 July:
LCDR . , ?VW 17 LSO	Pilot	In forward cockpit; starting starboard engine.
LTJG . VF-11	RIO	Preflighting rear cockpit.
AMH1 . VF-11	Line Supervisor	Standing by to assist plane captain.
AO1 . VF-11	Team leader of a VF-11 conventional weapons loading team	Awaiting signal from pilot to check Sidewinder.
AO2 . VF-11	Member, VF-11 weapons loading team.	Inspecting load, having homed port TER-7 and plugged in LAU-10s.
AO3 . F-11	Member, VF-11 weapons loading team.	Had just actuated the home/step switch on port TER-7. His arm was on port inboard LAU-10/A waiting for WILSON.
AOAA . '-11	Member, VF-11 weapons loading team.	Preparing to make stray voltage check on starboard rocket harnesses.
AN . '-1 Div	Driver of starting tractor.	Awaiting completion of engine start.
AMSAN . VF-11 (Deceased)	Plane Captain #110	Believed to be on starboard side of #110.
AN . 1 Div	Tractor Driver	Removing tow bar from #110.

27. That at 1051H, LCDR ___ __. ___ . USN, pilot of F-4 #110 had started his starboard engine and was switching from external to internal electrical power, preparatory to starting his port engine.

All redactions are B-6.

28. That F-4 #110 was loaded with ordnance on external stores stations as follows:

a. Station 2 (port inboard) - One AIM-9B (Sidewinder) missile. Three LAU-10/A launchers with four ZUNIs each.

b. Station 3 (port aft fuselage) - One AIM-7E (Sparrow III) missile.

c. Station 5 (centerline) - One full 600 gallon external fuel tank.

d. Station 6 (starboard forward fuselage) - Same as Station 3.

e. Station 8 (starboard inboard) - Same as Station 2.

29. That each ZUNI on F-4 #110 had a MK24 warhead, a M414A1 VT nose fuze and a MK191 Mod 1 base fuze.

30. That after pilot   had manned F-4 #110, tractor driver   rove up and parked his MD3A tractor forward of and close to the starboard wing of F-4 #110.

31. That   onnected up the starting unit and electrical power from his tractor to F-4 #110.

32. That on signal from pilot   provided starting air and electrical power to F-4 #110.

33. That at 1051H,   was sitting on his tractor, watching pilot   in F-4 #110 and observing   go through the starting procedures.

34. That line supervisor   who was checking the starting progress and preparation for launch of VF-11 F-4's, was standing generally in front of F-4 #110 at 1051H. He was not aware of the location of   o1   or what they were doing.

35. That after pilot   d manned F-4 #110 and after   had supplied external electrical power to the F-4,   began conducting stray voltage checks on #110.

All redactions are B-6.

36. That   had checked the LAU-17/A pylon and the three rocket harnesses on the port side of F-4 #110 for stray voltage using VF-11 procedures. No measurable voltage was detected.

37. That having completed the stray voltage checks on the port side of F-4 #110, HOWARD moved to the starboard side of #110 and at 1051H was preparing to conduct stray voltage checks on the starboard side.

38. That   along with   ',   and   , had helped load VF-11 F-4's with ordnance the morning of 29 July 1967 in preparation for the scheduled 1100H launch.

39. That after the VF-11 F-4's were loaded,   proceeded with   aft on the flight deck to F-4 #110 to ready the loaded ordnance for flight.

40. That after   had conducted the stray voltage checks on the port side of F-4 #110 as previously described,   checked the port Sparrow and pulled the pin from the port Sparrow launcher.

41. That   then visually checked the three LAU-10/A shorting devices to assure that they were on "safe"; checked the TER-7 electrical safety pin visually to assure that it was still installed and "in"; homed the TER-7 and checked that the green light was "on".

42. That having accomplished the preceding steps,   plugged in the three port rocket harnesses.

43. That at 1051H, after plugging in the port rocket harnesses,   was kneeling about two feet from the port TER-7 looking the aircraft over to be sure everything had been done right. He had spent about one minute at this task.

44. That   was following, assisting and observing   is he went through the preceeding steps to prepare F-4 #110 for flight.

All redactions are B-6

31

45. That ___ asked ___ f he had homed the port TER-7 on F-4 #110 and, not understanding the reply, ___ tuated the home-step switch on the TER-7. At this time (1051H) he was kneeling with his arm over the inboard LAU-10/A launcher.

46. That at 1051H, ___ was standing forward of F-4 #110 awaiting a signal from pilot ___ to begin the check of a Sidewinder.

47. That at 1051H, ___ was in front of F-4 #110 engaged in removing the tow bar from #110.

48. That at 1051H, plane captain ___ deceased) was in the vicinity of F-4 #110; however, his exact position is not known.

All redactions are B-6.

32

FINDINGS OF FACT

SECTION II

SEQUENCE OF PERTINENT EVENTS

49. That at 10-51-21H 29 July 1967, a single ZUNI rocket fired from one of the three LAU-10/A launchers installed on the port inboard wing station of F-4 #110 which was then spotted on the extreme starboard quarter of the flight deck, headed inboard at approximately a 45° angle to the ship's head.

50. That the ZUNI crossed the flight deck and struck A-4 #405, spotted on the port side of the flight deck, some 100 feet distant.

51. That both B-6 nd B-6 mneeling adjacent to the inboard pylon, under the port wing of F-4 #110, received minor burns from the ZUNI as it fired.

52. That B-6 's injuries were almost identical to those received by an ordnanceman aboard USS HANCOCK who was in a position equivalent to B-6 s when a ZUNI accidentally fired from a B-6 '-8 aircraft on 19 May 1967. See enclosure (143).

53. That the ZUNI passed very close to B-6 AFCM, B-6 USN, VA-46, who was standing in line between F-4 #110 and A-4 #405. The blast therefrom knocked B-6 vn.

54. That the reflection of the flash created by the ZUNI traveling across the deck was captured and recorded by the PLAT camera, installed on the 08 level of the island. (This was later verified by the Board in a demonstration conducted on 15 August 1967, the details of which are set forth in enclosures (137) and (138). The PLAT camera operator then seeing the flash

aft, swung his camera to the after portion of the flight deck and locked it in place. He later abandoned the station when the island was sprayed with shrapnel. From this time on, events on the after portion of the flight deck were recorded by the PLAT camera. The clock face showing on the PLAT film on the forenoon of 29 July was synchronized on -8 (HOTEL) zone time.

55. That numerous witnesses either saw or heard what looked or sounded like a rocket firing across the deck from starboard to port.

56. That the ZUNI broke up on impact with A-4 #405.

57. That the full 400 gallon tank on A-4 #405 was ruptured by the ZUNI, spreading the JP-5 fuel under A-4s #405 and #416, igniting the fuel.

58. That the fuel which poured onto the deck was quickly ignited by numerous fragments of burning rocket propellant.

59. That ⸺ B — G ⸺ USN, VA-46, who was standing near A-4 #405, was struck by the ZUNI or by shrapnel;

60. That B-6 7, ADJ3, B-G USN, VA-46, who was standing some 80 to 100 feet forward of A-4 #405, facing aft, was struck by a fragment from the ZUNI M414A1 VT nose fuze. The fragment was traveling essentially horizontally

61. That a fragment, type unknown, punctured the centerline external fuel tank of A-4 #310, which was just aft of the jet blast deflector of catapult #3, allowing fuel to pour onto the deck.

62. That fuel and fire instantaneously spread under A-4s #405 and #416.

63. That the burning fuel was then rapidly spread aft and fanned by 32 knots of wind over the deck from 350° relative and by the exhausts of at least three jets spotted immediately forward.

64. That at 1052H fire quarters was sounded.

65. That general quarters was sounded at 1053H, immediately following fire quarters.

66. That condition ZEBRA was set throughout the ship at 1059H; however, closure was not made along the routes being used to move injured personnel.

67. That personnel quickly manned high capacity fog foam (HCFF) and salt water fire hoses in the catwalks.

68. That because of the rapid spread of the fire, HCFF and salt water hoses abaft frame 190 on the port side were engulfed in flames and could not be used.

69. That at least one of the AN-M65 1000# bombs on A-4 #405 dropped or was ejected to the deck, due to short circuiting or to the direct impact of the ZUNI.

70. That a bomb which fell to the deck from A-4 #405 came to rest in a pool of burning JP-5 near the nose of A-4 #416, where it was rapidly heated.

71. That there was a longitudinal split in the AN-M65 1000# bomb which was lying on deck. The split bomb was observed to be burning brightly.

72. That 54 seconds after the fire began, ⎯⎯⎯⎯⎯, ABHC, ⎯⎯ V-1 Division, who was leading petty officer of the ship's crash and salvage crew, arrived at the scene and attempted to extinguish the fire in the vicinity of the above mentioned bomb with a hand held PKP (purple "K" powder) extinguisher.

73. That at approximately one minute 20 seconds after initiation of the flight deck fire the first hose played salt water on the forward boundry of the fire.

74. That at one minute 34 seconds after the start of the fire, a bomb exploded on the flight deck with approximately 35 personnel in close proximity, including two hose crews (HCFF station 9 and Salt Water Station 15).

75. That this first bomb explosion decimated the hose teams in the vicinity, causing nine casualties to the ship's crash and salvage crew. There were also eighteen casualties to other on-the-scene fire fighters.

76. That this first explosion spread the fire to the group of three A-4 aircraft spotted across the stern.

77. That immediately after the first explosion, several other hose teams continued to lead their hoses toward the fire.

78. That nine seconds after the first bomb exploded, a second bomb exploded at the after end of the flight deck with even more violence than the first.

79. That this second major explosion hurled bodies and debris as far as the bow and extended the fire along the seven F-4's and toward the three RA-5C's on the starboard side abaft the island.

80. That seven major explosions occurred at times indicated in enclosure (135).

81. That effective fire fighting efforts on the flight deck were interrupted after the second major explosion for approximately five minutes until major explosions had subsided.

82. That during the period of major explosions, activity continued on the forward part of the flight deck and behind the shelter of the island in the form of assistance to the injured, jettisoning of ordnance, and break out of hoses to be led aft.

83. That fuel, spilled from the stricken aircraft, ran over the ship's sides and stern setting fires on the sides, sponsons, fantail, and in hangar bay 3.

84. That a total of approximately 40,000 gallons of JP-5 fuel was aboard the burning aircraft on the flight deck and this fuel fed the flames.

85. That the force of bombs exploding on the flight deck penetrated to hangar bay 3, starting fires on the 03, 02 and 01 decks aft.

86. That the force of bombs which exploded on the flight deck; in penetrating to the 03 deck aft; killed some 50 sleeping night check crew personnel, and others, for a total of 91 killed in the after areas of the ship. (See enclosure (317)).

87. That the sprinkler system in hangar bay 3 was activated in time to prevent spread of fire in hangar bay 3. Fog foam monitors were shortly thereafter activated in hangar bay 3.

88. That fog foam and sprinklers effectively prevented the spread of fire further forward on the hangar deck.

89. That shortly after the fire commenced, personnel began to jettison over the side or to strike below all exposed explosive ordnance on the flight, hangar, and second decks.

90. That after the seven major explosions ceased, aircraft forward of the RA-5C's were moved forward and effective flight deck fire fighting commenced; a fire boundary was established on the flight deck athwartship, at approximately frame 165.

91. That after the major explosions ceased, 3 RA-5C's and 1 A-4E were jettisoned over the side because they were on fire, or were leaking fuel, which could have further spread the fire.

101. That at least 20 men jumped, were blasted, knocked or fell overboard and were recovered, or are missing. (9 were recovered by helos from accompanying ships; 6 were recovered by USS MACKENZIE (DD-836); 1 was recovered by means unknown; and 4 are believed missing.)

102. That 134 personnel are dead or missing as a result of the fire and explosions.

103. That 161 personnel were injured as a result of the fire and explosions.

104. That as of 15 September 1967, the known estimated cost of damage, exclusive of aircraft damage, and costs of the equipments and systems under the cognizance of Naval Air Systems Command, is $72,103,000.

FINDINGS OF FACT

SECTION III

CIRCUMSTANCES BEARING ON INITIATION OF THE FIRE

105. That LCDR ⁊⬦ CVW-17 Senior LSO, was scheduled for a flak suppression/target CAP mission with VF-11 on 29 July 1967 flying F-4B BUNO 153061, modex #110, with launch scheduled for 1100H.

106. That F-4 #110 was flown previously on the morning of 29 July, landing at 0902H. One MK82 500# bomb loaded on TER station 3 on external stores station 8 (starboard wing) did not release and was returned to the ship.

107. That as a result of this discrepancy, an armament release and control system check of station 3 of the TER-7 on external stores station 8 was conducted. The discrepancy was believed to be due to a faulty cartridge in TER station 3.

108. That LCDR ⁊⬦ was told some 5 minutes before manning his aircraft for the scheduled 1100H launch that he would be carrying ZUNIs rather than 2.75-inch rockets.

109. That CDR ⬦⁊ , CO VF-11, who was Flight Leader of the VF-11 scheduled 1100H flight, briefed LCDR ⬦⬦ as to the changes in flight procedures which would be required because of the loading change to ZUNIs. He emphasized particularly sight settings and whether or not to bring the empty launchers back to the ship.

110. That no mention was made during the above briefing as to checking shorting devices on the LAU-10/A launchers during the pre-flight inspection of the aircraft.

111. That F-4 #110 was spotted on the extreme starboard quarter of the flight deck, headed inboard, at approximately a 45 degree angle to the ship's head.

112. That LCDR l ⟨illegible⟩ in his pre-flight inspection of F-4 #110, noted particularly the following:

    a. LAU-10/A launchers - Frangible firings (forward) -- secure.

                               - Selector switches -- ripple.

    b. TER-7 - Electrical and mechanical safety pins in; ordnance secure; rocket harnesses unplugged.

    c. Sidewinder - Safety pins in; no cover on port Sidewinder.

113. That neither LCDR ⟨illegible⟩ nor LTJG l ⟨illegible⟩ his RIO, checked the shorting devices on the LAU-10/A launchers loaded on F-4 #110.

114. That upon manning F-4 #110, LCDR ⟨illegible⟩ set his armament switches in the following positions:

    a. Missile control pannel:

        (1) Power switch -- Off

        (2) Arm-safe switch -- Safe

        (3) Select switch -- Radar

        (4) Interlock switch -- Out

    b. Bomb control panel:

        (1) Bomb control switch -- Direct

        (2) LABS Mode switch -- Instantaneous Over the Shoulder

    c. Weapons control panel:

        (1) Weapons select switch -- Bull Pup

        (2) Weapons switch -- Conventional Off, Nuclear On

        (3) Station selector -- Off

        (4) Bomb arming switch -- Safe

115. That LCDR ⟨illegible⟩ did not check to see that the armament safety override switch was in the safe (out) position.

116. That the change in loading of F-4 #110 from 2.75 inch rockets to ZUNIs was made sufficiently far ahead of loading time for the scheduled 1100H launch that the ordnance crews were not unduly rushed.

117. That the jet starter tractor driven by AN ⟨redacted⟩ , V-1 Division, was positioned forward of the starboard wing of F-4 #110 rather than forward of the port wing, as was normal, because of the close proximity of A-4 #410 on the port side of F-4 #110.

118. That the jet starter tractor which had started A-4 #410 had been moved prior to the start of the fire.

119. That LCDR ⟨redacted⟩ started his starboard engine, deselected air at 45% RPM, and, as the engine stabilized at 65%, selected both generators.

120. That as LCDR ⟨redacted⟩ selected both generators, a mild explosion shook his aircraft, appearing to LCDR ⟨redacted⟩ " to come from the starboard side of F-4 #110.

121. That LCDR ⟨redacted⟩ , looking up at the time of the mild explosion saw an object flying across the deck. It appeared to him to be small, issuing a yellow-orange flame. The object struck an A-4 parked on the port side of the ship directly in front of LCDR ⟨redacted⟩ airplane.

122. That the A-4 struck was #405.

123. That LCDR ⟨redacted⟩ s immediate impression was that one of the ZUNI rockets had fired from his aircraft when he switched from external to internal power.

124. That LCDR ⟨redacted⟩ then rechecked his cockpit armament switches after the object hit the A-4, and found all switches to be in the positions previously described.

125. That as a general safety measure, each rocket harness is checked by a voltmeter to assure that there is no stray voltage in the circuit which would fire a rocket when the harness is connected.

126. That it is current practice aboard attack carriers operating on Yankee Station conducting combat strikes, to plug in rocket launchers in the initial spot (in the pack) prior to being taxied to the catapults.

127. That the practice of plugging in rocket launchers prior to reaching the catapults, though not documented, was well known and tacitly approved.

128. That on the morning of 29 July 1967, B.b was conducting the stray voltage check on F-4 #110 in preparation for plugging in the rocket launchers, which was to be accomplished by WILSON.

129. That B.b had conducted a stray voltage check on the port TER-7 of F-4 #110 and at 1051H had proceeded to the starboard side of #110 preparatory to conducting a stray voltage check on the starboard TER-7.

130. That the stray voltage check conducted by B.b on the port TER-7 of F-4 #110 was as described in Section VI hereof. The procedures he used were those employed in VF-11 and were undocumented.

131. That the above mentioned stray voltage check on the port side of F-4 #110 was, however, conducted while the aircraft engines were being started and while F-4 #110 was still on external electrical power; hence, before the aircraft electrical system had stabilized.

132. That no official documentation from any source, was known to exist prior to 29 July, as to precisely when, how, or why this stray voltage check should be conducted prior to plugging in LAU-10/A ZUNI launchers on the TER-7 installed on F-4B aircraft.

133. That no official documentation from any source was available prior to 29 July which specified that the stray voltage check and subsequent rocket plug in must be conducted after an F-4B aircraft is started and has switched to internal electrical power.

43

134. That measurable transient voltages occur in the F-4B aircraft when switch is made from external to internal electrical power.

135. That VF-11 had two conventional weapons loading teams; one under AO1 LICHTSEY and the other under AO1 ROGERS.

136. That the VF-11 ordnance CPO, ₿⁻ᵇ ₹, and the two VF-11 conventional weapons loading team leaders, ↑ᵧᵇ and ₿ᵢₒ , considered it safe to conduct stray voltage checks and to plug in rockets while the aircraft was still on external electrical power.

137. That Commander ₿·ᵇ , CO of VF-11, considered it safe to plug in rockets while the F-4 was still on external electrical power but stated that it was squadron policy not to do so.

138. That VF-11 squadron policy and procedures with regard to conducting stray voltage checks and plugging in rockets were not documented.

139. That the stray voltage check as normally conducted by VF-11 ordnancemen measured only the voltage that might be present in the rocket harness and the short wire inside the TER-7 which leads to the open select rocket solenoid switch therein.

140. That the stray voltage check as conducted by VF-11 did not check the firing circuitry upstream of the select rocket solenoid switch in the TER-7.

141. That after ₿·ᵇ onducted his stray voltage check of the port TER-7 of F-4 #110, ₿·ᵇ visually checked the position of the TER-7 electrical safety pin, checked the three LAU-10/A shorting devices, homed the TER-7, checked the green light which indicated that the TER-7 was homed, and then plugged in the three LAU-10/A ZUNI launchers attached thereto. The preceding actions were accomplished while F-4 #110 was still on external power.

142. That having plugged in the port LAU-10/A's, B/b spent about one minute checking the port TER-7 and LAU-10/A's of F-4 #110 and was so engaged when, at 10-51-21H, the ZUNI fired from that aircraft.

143. That on the morning of 29 July, B/b was following, observing and helping Ty/b ready the TER-7 and LAU-10/A's under the port wing of F-4 #110.

144. That B/b asked D/b if he had homed the port TER-7 of F-4 #110.

145. That B/b was unable to understand R/b s reply so he actuated the home-step switch of the port TER-7 of F-4 #110. This occurred after the three LAU-10/A launchers had been plugged in by R/b

146. That after i Ty/b actuated the home-step switch, the ZUNI fired from F-4 #110.

147. That a ZUNI can fire from an F-4B configured as was F-4 #110 preparatory to the scheduled 1100H launch on 29 July, even though all cockpit armament switches are in the prescribed (safe) positions, if the following conditions exist simultaneously:

a. If current is present at the TER-7 in the stepping and firing circuit.

b. If the select rockets solenoid switch in the TER-7 is closed (in the firing position).

c. If the TER-7 electrical safety switch is in the armed position.

d. If the TER-7 rocket harness is connected to the LAU-10/A.

e. If the LAU-10/A shorting device is not grounding the launcher circuitry.

148. That the CA42282 pylon electrical disconnect (commonly called the 101 pin connector), which connects the F-4 airframe electrical system to the LAU-17/A pylon, has a history of shorting due to the presence of unwanted moisture contamination and subsequent corrosion.

149. That shorting of pin #23 within the CA42282 pylon electrical disconnect can provide firing and stepping current to the TER-7.

150. That shorting of pin #73 within the CA42282 pylon electrical disconnect can provide current to close the select rockets solenoid switch in the TER-7.

151. That as discussed in detail in ensuing findings, an armament electrical system check conducted on the starboard TER-7 of F-4B BUNO 152278 and subsequent can damage the electrical safety switch in the port TER-7 model -527 due to direct grounding of high magnitude current, thereby causing this switch to fail.

152. That the LAU-10/A shorting device is unreliable and, even though apparently in the safe position, can allow firing current to pass from the rocket harness through the LAU-10/A to the ZUNI.

153. That the model -527 TER-7 racks aboard ⟨⟩⟨⟩ are not wired as indicated in NAVAIR 11-75A-40 of 15 August 1966. They are actually wired with the safety switch shorted to ground, instead of in series as shown in the schematic.

154. That the change dated 15 April 1967 to NAVAIR 11-75A-40 of 15 August 1966 indicating the correct wiring of the -527 TER-7 was not received aboard FORRESTAL until after 29 July 1967.

155. That no one in FORRESTAL was aware of the above wiring change and its impact on F-4B firing circuitry.

156. That since the wiring of F-4 #110 (BUNO 153061) was such that power was directed to both inboard and outboard stations in pairs, the firing circuit continuity check made on the TER-7 on Station 8 (starboard wing) would have sent 28 volt DC current to Station 2 (port wing). Since the -527 TER-7 safety switch is shorted to ground, this test

could have damaged the safety switch of the TER-7 on Station 2, causing it to malfunction.

157. That the -527 TER-7 safety switch can also be damaged by a high magnitude current through a shorted #23 pin in the CA42282 pylon electrical disconnect.

158. That during the period 25-29 July, F-4 aircraft had arrived at the catapults with one or more TER-7 electrical safety pins missing.

159. That TER electrical safety pins had been found adrift on the flight deck which had presumably been pulled out by the wind and jet blast acting on the long red warning flag attached thereto.

160. That the ordnance crew working around F-4 #110 at 1050H on 29 July 1967 was not one of the two previously mentioned conventional weapons loading teams that were headed by AO1   ' and AO1   , but was composed of members of both teams.

161. That team leader   was preparing to check a Sidewinder on F-4 #110 at 1051H on 29 July but was not in charge of the crew composed of   and   then working on #110.

162. That team leader   was in charge of the crew working on F-4 #110 at 1051H on 29 July but he,   was not present at the aircraft to observe the operations.

163. That at no time were hand signals given to pilot   to indicate to him that ordnancemen were at work on his aircraft.

164. That a VF-11 squadron policy, newly established, prior to 29 July, eliminated the safety ordnance petty officer in the pack and gave the responsibility to the plane captain of giving all hand signals to the pilot.

165. That as a result of this policy, no provision was made for informing the pilot that ordnancemen were working under his aircraft.

All redactions are B-6

FINDINGS OF FACT

SECTION IV

SHIP ORGANIZATION, RESPONSIBILITIES, TRAINING AND OPERATING PROCEDURES

166. That the following officers attached to USS FORRESTAL (CVA-59)
were made parties to the investigation into the circumstances surrounding
the fire on FORRESTAL on 29 July 1967, because of their responsibilities
at that time as set forth below:

a. Captain ___ USN, ___ Commanding Officer,
USS FORRESTAL (CVA-59) was, as set forth in U. S. Navy Regulations,
1948, Chapter 7, charged with absolute responsibility for the safety,
well being, and efficiency of his command.

b. Commander ___ D, USN, ___ Engineering
Officer, USS FORRESTAL (CVA-59) was assigned the following duties,
among others, by NAVAIRLANT/NAVAIRPAC CV SHIP INST 5400.1:

(1) Supervise and direct the control of damage in the ship
caused by casualty, flooding, or fire.

(2) Supervise and direct the fighting of fires in the ship,
except aircraft fires.

c. Lieutenant Commander J___ ___ USNR, ___ Damage
Control Assistant, USS FORRESTAL (CVA-59) was assigned to the following
duties, among others, by NAVAIRLANT/NAVAIRPAC CV ENG INST 5400.1:

(1) Direct damage control and fire fighting operations during
emergencies or drills.

(2) Train ship's personnel in damage control including fire
fighting and emergency repairs.

(3) Coordinate and supervise the routine testing of fire
fighting equipment and other damage control equipment by all divisions.

d. CWO2 ~~B-℀~~ , USN, ~~B.6~~ Fire Marshall, USS FORRESTAL (CVA-59) was assigned the following duties, among others, by NAVAIRLANT/NAVAIRPAC CV ENG INST 5400.1:

(1) As Fire Marshall, conduct inspections, analyze conditions, make recommendations, conduct training, take corrective action, and perform such other duties as may be assigned for fire prevention, fire protection, and fire fighting on board the vessel.

e. Commander ~~B.℀~~ USN, ~~B.6~~ . Air Officer, USS FORRESTAL (CVA-59) was assigned the following duties, among others, by NAVAIRLANT/NAVAIRPAC CV SHIP INST 5400.1:

(1) Supervise and direct aircraft fire fighting, crash removal, and salvage operations, including the organization and training of personnel concerned with these operations.

f. Lieutenant Commander ~~B-6~~ USN, ~~B-6~~ Hangar Deck Officer, USS FORRESTAL (CVA-59) was assigned the following duties, among others, by NAVAIRLANT/NAVAIRPAC CV AIR INST 5400.1:

(1) Maintain assigned hangar deck fire fighting equipment, including CO_2, Foamite, and salt water outlets, and supervise the fighting of hangar deck fires until properly relieved.

g. Lieutenant ~~B-6~~ , USN, ~~B-6~~ . Flight Deck Officer, USS FORRESTAL (CVA-59) was assigned the following duties, among others, by NAVAIRLANT/NAVAIRPAC CV AIR INST 5400.1:

(1) Supervise flight deck crash crews and fire parties during the removal of plane crashes, plane salvage, pilot rescue, plane jettisoning, or flight deck fires.

h. WO1 ~~R b~~ BN, ~~T b~~ Aviation Ordnance Gunner, USS FORRESTAL (CVA-59) was assigned the following duties, among others, by NAVAIRLANT/NAVAIRPAC CV GUN INST 5400.1:

(1) Supervise, as required, the arming and de-arming of embarked aircraft insuring the observance of safety precautions and proper procedures.

(2) Insure the strict observance of all safety procedures and regulations relative to the handling of ordnance materials.

167. That Captain ~~R b~~ took command of USS FORRESTAL on 7 May 1966 at Norfolk Naval Shipyard, Portsmouth, Virginia.

168. That FORRESTAL, having returned from an extended Mediterranean deployment, entered the NNSY on 15 April 1966 with an experienced crew.

169. That FORRESTAL underwent extensive overhaul at the NNSY, from 15 April 1966 to 23 January 1967.

170. That FORRESTAL's employment subsequent to completing overhaul 23 January 1967 until arrival on Yankee Station, 24 July 1967, was as follows:

a. NNSY to Virginia Capes Operating Area: Individual Ship's Exercise, 23 to 27 January 1967.

b. At anchorage, Norfolk, Virginia, 27 to 30 January 1967.

c. Norfolk Naval Station to Virginia Capes Operating Area: INSURV Trials, 30 January through 1 February 1967. Trials continued at Pier 12 until 3 February.

d. Norfolk Naval Station, Pier 12: On-load of equipment and aircraft, 1 February to 14 February 1967.

e. Enroute Norfolk to Guantanamo Bay, Cuba: 14 to 17 February 1967.

f. At Guantanamo: Refresher Training, 17 February to 18 March 1967.

g. Enroute Guantanamo to Atlantic Fleet Weapons Range, 18 to 19 March 1967.

h. Type training at AFWR, 20 to 24 March 1967.

i. Enroute AFWR to Norfolk, 25 to 27 March 1967.

j. Restricted availability at NNSY, 28 March to 11 April 1967.

k. Enroute Norfolk to AFWR, 12 to 14 April 1967.

l. Type training at AFWR, 14 April to 3 May 1967 participating in Exercise CLOVE HITCH III from 21 to 24 April and NTPI 3 to 5 May 1967.

m. Enroute from AFWR to Norfolk, 3 to 6 May 1967.

n. At Pier 12/anchorage, Norfolk, 6 to 11 May 1967.

o. Carrier qualifications and ACLS trials in the Virginia Capes Operating Area, 11 to 18 May 1967.

p. At anchorage, Norfolk, 19 May 1967.

q. Family day cruise, 20 May 1967.

r. Prepared for overseas movement at Norfolk, 21 May to 5 June 1967.

s. Departed Norfolk for WESTPAC, 6 June 1967.

t. Type training at AFWR, 9 to 13 June 1967.

u. Operational Readiness Inspection, 14 to 16 June 1967.

v. Enroute Rio de Janeiro, 16 to 23 June 1967.

w. Port visit Rio de Janeiro, 23 to 26 June 1967.

x. Enroute WESTPAC, 26 June to 18 July 1967.

y. Chopped to COMSEVENTHFLT, 8 July 1967.

z. Arrived Subic Bay, R. P., 18 July 1967.

aa. Enroute Subic Bay to Yankee Station, 22 to 23 July 1967.

bb. Arrived on Yankee Station, 24 July 1967.

171. That since completing overhaul on 23 January 1967, FORRESTAL has had an average on-board count of 140 officers and 2775 men belonging to ship's company. FORRESTAL allowance is 126 officers; 2508 men.

51

172. That upon completion of post repair trials an INSURV inspection of FORRESTAL, as requested by the CO, was conducted during the period 30 January to 3 February 1967 to identify material discrepancies. Enclosure (386) summarizes the overall condition of damage control within the ship as barely satisfactory.

173. That during the period 14 to 17 February 1967 enroute to Guantanamo Bay, Cuba, FORRESTAL conducted training preparatory to refresher training.

174. That during the period 17 February to 18 March 1967, FORRESTAL underwent refresher training at Guantanamo. Upon completion of this training COMFLETRAGRU Guantanamo conducted an Operational Readiness Inspection and assigned the following grades for Damage Control.

    a. Battle Problem.

        (1) Setting Material Condition   – Unsatisfactory

        (2) Damage Control   – Excellent

        (3) NBC   – Satisfactory

        (4) Medical   – Excellent

    b. Operational Exercises.

        (1) Damage Control   – Good

        (2) Fire Drill (In port)   – Good

        (3) Fire at Sea   – Excellent

        (4) Rescue and Assistance Drill   – Good

        (5) Collision Drill   – Satisfactory

    c. Specialty Phase.

        (1) HCFF Injector Sections   – Excellent

Complete listing of all grades assigned by COMFLETRAGRU Guantanamo is attached as enclosure (337).

175. That USS FORRESTAL departed Norfolk, Virginia, for deployment to WESTPAC on 6 June 1967.

176. That during the period 14 to 16 June 1967, FORRESTAL with embarked CVW-17 participated in its Operational Readiness Inspection (ORI), conducted by COMCARDIV TWO, at the AFWR. Pertinent grades and comments were:

    a. The 5MC could not be heard in some areas of the flight deck.

    b. Nozzles were not installed on hangar deck fire hoses.

    c. Overall performance of damage control parties above the main deck was weak.

    d. Conventional Weapons Loading was graded OUTSTANDING. Crews were quick, alert, and adaptable to changing situations.

    e. Flight Deck Fires was graded OUTSTANDING. Quick and effective response to extinguish a simulated tail pipe fire resulted in a grade of 99 for this exercise.

A complete listing of all grades assigned is attached as enclosure (388).

177. That during the transit from AFWR to Cubi Point, R. P., FORRESTAL conducted extensive training in damage control and fire fighting. These efforts are depicted in enclosure (89).

178. That FORRESTAL's average on board count during overhaul was 2,796 men.

179. That during the overhaul period, 1,496 enlisted men were transferred from FORRESTAL, representing 54 percent of the average on board count.

180. That during the overhaul period, FORRESTAL received on board 1,500 enlisted men from other ships and stations.

181. That during the overhaul period, 261 supervisory personnel (E-5 through E-9) were transferred from FORRESTAL; 192 supervisory personnel were received.

182. That since overhaul, from January through July 1967, FORRESTAL received 1,039 enlisted men and transferred 988 for a net gain of 51 enlisted men.

183. That during the period from January through July 1967, FORRESTAL received 241 supervisory personnel, and transferred 207 for a net gain of 34 supervisory personnel.

184. That during the period following refresher training and prior to deployment to WESTPAC, 19 March to 6 June 1967, 717 personnel were transferred from FORRESTAL, of which 151 were supervisory personnel.

185. That, in summary, the following enlisted personnel transfers occurred:

    a. Shipyard, 15 April 1966 to 23 January 1967 – total: 1,496

                                               Supervisory: 261

    b. Post-shipyard, 23 January to 6 June 1967 – total: 951

                                               Supervisory: 197

186. That the following percentage of on board count versus allowance of supervisory personnel existed for the following selected rates: (Data for all ratings contained in enclosure (89)).

		% JAN 67	% JUL 67
a.	SF	78	39
b.	DC	79	57
c.	AO	46	103

187. That subsequent to overhaul, during the period January through July 1967, FORRESTAL conducted general quarters training periods as shown below:

a.	JAN	6	e.	MAY	9
b.	FEB	11	f.	JUN	10
c.	MAR	14	g.	JUL	4
d.	APR	3		Total:	57

188. That subsequent to overhaul, during the period January to July 1967, FORRESTAL conducted training in fire fighting evolutions as shown below:

MONTH		DRILL	ACTUAL FIRES	TOTAL
a.	JAN	1	4	5
b.	FEB	12	6	18
c.	MAR	21	5	26
d.	APR	11	6	17
e.	MAY	11	2	13
f.	JUN	6	9	15
g.	JUL	1	10	11
		63	42	105

189. That during the period from 15 April 1966, when the ship entered NNSY, until 29 July 1967, 1,332 officers and men graduated from fire fighting schools of at least 2 days duration; of these 1,332, 327 had been transferred by 29 July 1967.

- -

190. That responsibilities for the control and handling of conventional (and nuclear) weapons aboard FORRESTAL were documented in USS FORRESTAL Instruction 03510.2E of 2 May 1967.   Enclosure (89).

191. That by the preceding instruction, a Weapons Planning Board was established. A pertinent portion of the establishing directive is quoted: "within the parameters assigned by the Commanding Officer, make detailed plans for utilization of FORRESTAL as a weapon system in prosecuting all combat operations tactical or simulated, including attack, defense, deception weapons utilization, and movement".

192. That the Weapons Planning Board is comprised of the following officers:

    a. Operations Officer (chairman)

    b. Strike Operations Officer

    c. Air Wing Commander

    d. Air Wing Operations Officer

    e. Surface Systems Supervisor

    f. Weapons Officer

    g. Air Officer

    h. Air Operations Officer

193. That subordinate to the Weapons Planning Board is the Weapons Coordination Board which, "is a division level board instituted to work out all details necessary to execute the loading of ordnance as directed by the Weapons Planning Board".

194. That the Weapons Coordination Board is comprised of the following members:

    a. Strike Operations Officer (chairman)

    b. Aircraft Handling Officer

    c. Weapons Loading Director

    d. Nuclear Weapons Technical Supervisor

    e. Squadron Weapons Coordinators

    f. Air Wing Operations Officer

    g. Commanding Officer, Marine Detachment

195. That Captain ꟼꙅꙠ authorized the Weapons Coordination Board and Weapons Planning Board to establish procedures for ordnance handling which would vary as little as possible from Atlantic Fleet peacetime procedures, using known WESTPAC procedures as a guide.

196. That based on the foregoing authorization, ordnance handling procedures intended for use in WESTPAC deployment had been developed and were validated by inspectors during the Operational Readiness Inspection, 14-16 June 1967.

197. That on 29 June 1967 the Weapons Coordination Board, augmented by key members of the Weapons Planning Board, met to document those ordnance handling procedures which were in use and which had been validated during the Operational Readiness Inspection. These were the procedures to be used in WESTPAC.

198. That this 29 June meeting was also attended by experienced Warrant Officers and Chief Petty Officers from the squadrons, and, representatives from all ordnance handling divisions in the ship.

199. That CDR ʀ̶ᵇ. CO, VF-11, who was acting COMCVW-17, did not attend the 29 June meeting.

200. That the following doctrines and procedures applying to handling and arming LAU-10/A ZUNI launchers on the flight deck were embodied in the minutes of this meeting, as follows:

    a. Allow ordnance personnel to connect pigtails "in the pack", prior to taxi, leaving only safety pin removal on the cat.

    b. Strictest adherence to existing safety precautions provide the very minimum possibility of inadvertent firing.

    c. Ordnancemen must receive positive identification of awareness of pilot (and crewman, where applicable) that connection is being made, and "hands off" signal given and acknowledged.

    d. Stray voltage check must be scrupulously performed prior to connecting pigtail.

    e. Safety pin will not be removed prior to aircraft being positioned on the catapult, nor in the case of the LAU-10 will the arm lever be cocked.

201. That minutes of the 29 June meeting were taken and these minutes were read and agreed to by all attendees at the conclusion of the meeting.

202. That it was also stated at the conclusion of the 29 June meeting that a double spaced copy of the agreements reached would be distributed to squadrons and divisions for their correction of any ambiguous statement. Enclosure (83) is a copy of this document.

203. That on 8 July another draft of the agreements of the 29 June meeting was written, incorporating corrections which had been recommended. This draft was not distributed to either squadrons or divisions. Enclosure (84) is a copy of this document.

204. That in so far as CO, FORRESTAL, ship's Operations Officer, and COMCVW-17 were concerned, the procedures as used in the ORI relating to the ZUNI, remained in effect.

FINDINGS OF FACT

SECTION V

CARRIER AIR WING SEVENTEEN ORGANIZATION, RESPONSIBILITIES,
TRAINING, AND OPERATING PROCEDURES

205. That the following officers and men attached to Carrier Air Wing
17 were made parties to the investigation into the circumstances
surrounding the fire on FORRESTAL because of their responsibilities or
duties at that time as set forth below:

a. CDR . B-6 , Jr., USN, B-6 , Commander, Carrier
Air Wing 17, embarked in USS FORRESTAL (CVA-59), was responsible to the
Commanding Officer, USS FORRESTAL (CVA-59), for, among other things:

(1) The coordination and supervision of all activities of the
squadrons and detachments in the Air Wing.

(2) The actual loading of conventional munitions on Air Wing 17
aircraft. This included the responsibility for the observance of ordnance
safety precautions by Air Wing personnel with regard to handling,
dearming and maintenance of aircraft ordnance and equipment (FORRINST
03510.2E of 2 May 1967).

b. LCDR B-6 , USN, B-6 was the Carrier Air
Wing 17 Senior LSO (Landing Signal Officer). At 1051H on the morning
of 29 July 1967, he was in the forward cockpit of F-4B #110 of VF-11,
starting the starboard engine in preparation for a scheduled launch.

c. LT B-6 , USN, B-6 , Carrier Air Wing 17
Ordnance Officer, was responsible to the Commander Carrier Air Wing 17
for the performance of the following duties:

(1) Advise COMCVW-17 in ordnance matters.

(2) Act as point of liaison between ship/squadrons/COMCVW-17
on ordnance matters, with particular regard to safety, and timely load-out
of squadron aircraft.

(3) Perform quality control checks of ordnance loading of CVW-17
aircraft.

(4) Make spot checks of general flight deck ordnance evolutions.

59

d. CDR ___B-6___ USN, B-6 Commanding Officer, Fighter Squadron 11 (VF-11), attached to Carrier Air Wing 17 in USS FORRESTAL (CVA-59), was responsible to the Commander, Carrier Air Wing 17 for, among other things:

(1) The overall operational readiness of his squadron.

(2) The effective preparation of his squadron for all missions.

(3) The actual loading of conventional munitions on VF-11 aircraft. This included the responsibility for the observance of ordnance safety precautions by VF-11 personnel with regard to handling, dearming and maintenance of aircraft ordnance and equipment (FORRINST 03510.2E of 2 May 1967).

e. LCDR B-6 , Jr., USN, B-6 was Aircraft Maintenance Officer, VF-11. As Department Head, he was responsible to the Commanding Officer, VF-11, for the accomplishment of the mission of his department. The Avionics/Weapons Division and the Weapons Branch were in the Aircraft Maintenance Department.

f. LT B-6 , USN, B-6 was Avionics/Weapons Division Officer, Aircraft Maintenance Department, VF-11. He was responsible to the Aircraft Maintenance Officer for the performance of the tasks assigned his division with maximum effectiveness. His duties included:

(1) Ensure compliance, by all personnel of his division, with all safety precautions issued by higher authority.

(2) Prepare and submit for promulgation such additional safety precautions as required.

g. WO1 B-6 , USN, B-6 was Assistant Avionics/ Weapons Division Officer, Aircraft Maintenance Department, VF-11. He assisted the Avionics/Weapons Division Officer in the performance of all of his assigned duties.

h. LTJG ⬚ Jr., USNR, ⬚ was the Weapons
Branch Officer, Avionics/Weapons Division, Aircraft Maintenance
Department, VF-11. He was responsible in accordance with paragraph
1044, chapter 10, U.S. Navy Regulations 1948, to the Avionics/Weapons
Division Officer for the proper performance of the duties assigned
to the Weapons Branch, including the training of the ordnancemen
attached to the Weapons Branch.

i. Chief Aviation Ordnanceman (AOC) ⬚ N, ⬚
Aviation Ordnance Supervisor, was the senior petty officer attached to
the Weapons Branch, Avionics/Weapons Division, Aircraft Maintenance
Department, VF-11. He assisted LTJG ROWAN in the supervision and
performance of all duties assigned to the Weapons Branch.

j. Aviation Ordnanceman First Class (AO1) ⬚ N
⬚ attached to the Weapons Branch, VF-11, was in charge of
one of the conventional weapons loading teams in the Weapons Branch,
which were organized in accordance with COMCVW-17 NOTICE 8000 of
17 June 1967. Members of his team were: AO2 ⬚, AO3 ⬚
AO3 ⬚, AOAN ⬚, AOAA ⬚ and AA ⬚. He was responsible
for proper and safe performance of duty by the members of his team.

k. Aviation Ordnanceman Second Class (AO2) ⬚ USN,
⬚ attached to the Weapons Branch, VF-11, was a member of
the conventional weapons loading team headed by AO1 ⬚ At
1051H on the morning of 29 July 1967, he was under the port wing
of F-4B #110 of VF-11. WILSON had completed homing the TER-7 on
external stores station 2 and had just plugged in the three LAU-10/A
rocket launchers loaded on the TER.

l. Aviation Ordnanceman Airman Apprentice (AOAA), ⬚ , USN
⬚ ttached to the Weapons Branch, VF-11, was a member of
the conventional weapons loading team headed by AO1 ⬚ At 1051H
on the morning of 29 July 1967, he was under the starboard wing of F-4B

#110 of VF-11, conducting stray voltage tests on the rocket harness of the TER-7 on external stores station 8, having just completed these tests on the TER on external stores station 2.

206. That CVW-17 (Carrier Air Wing Seventeen) was commissioned on 1 November 1966.

207. That CDR ꟼ〕 ⌐ᗷ assumed command of CVW-17 on 22 December 1966.

208. That LT. ꟼ-ᗷ had recent previous experience as an Air Wing Ordnance Officer with CVW-9 in WESTPAC in USS ENTERPRISE (CVA-65) and USS RANGER (CVA-61).

209. That CVW-17 was assigned as the USS FORRESTAL Air Wing for the 1967 Southeast Asia Deployment.

210. That the homeports of the squadrons and detachment of CVW-17 are as follows:

VF-11	NAS, Oceana
VF-74	NAS, Oceana
VA-65	NAS, Oceana
VA-46	NAS, Cecil Field
VA-106	NAS, Cecil Field
RVAH-11	NAS, Sanford
VAW-123	NAS, Norfolk
VAH-10 Det 59	NAS, Whidbey Island

211. That all of the squadrons and detachment of CVW-17 were never joined together for operations as a Carrier Air Wing until the departure of FORRESTAL from Norfolk, Virginia on 6 June 1967.

212. That during the transit to WESTPAC, all air wing personnel were trained in location of alternate escape routes from their living and berthing spaces and their work areas.

213. That training of the air wing in use of OBA's and firefighting was started during the REFTRA period and continued during the transit to WESTPAC, but was by no means completed.

214. That some 102 air wing Ensigns, LTJG's and Warrant Officers, who stood the aircraft integrity watches, were indoctrinated in the locations, activation and procedures for operation of hangar bay divisional doors, sprinkler systems, and the associated fire fighting equipment on the hangar and flight decks.

215. That all squadrons and detachment of CVW-17 were embarked in USS FORRESTAL for the ORI (Operational Readiness Inspection), conducted by COMCARDIV TWO 14-16 June 1967.

216. That the squadrons of CVW-17 received the following numerical and adjective grades on the ORI:

VF-11	90.8	EXCELLENT
VF-74	90.5	EXCELLENT
VA-106	92.7	EXCELLENT
VA-46	91.1	EXCELLENT
VA-65	96.4	OUTSTANDING
RVAH-11	90.4	EXCELLENT
VAW-123	88.5	EXCELLENT
VAH-10 Det 59	92.2	EXCELLENT

217. That CDR B-6 COMCVW-17, departed USS FORRESTAL (CVA-59), on 16 June 1967 on an advance liaison trip to the Far East to the carriers on Yankee Station. He returned to FORRESTAL on 15 July 1967.

218. That before departure, CDR B-6 discussed with Captain BELING the desirability of having a Weapons Coordination Board meeting during the ship's transit to WESTPAC, in order to conduct a detailed study and documentation of CVA-59/CVW-17 weapons handling procedures, as developed before and as used during the ORI.

219. That CDR D-6 R also directed LT B-6 to organize composite Air Wing catapult arming crews in order to comply with the ORI recommendation to reduce the numbers of personnel between the catapults.

220. That CDR B-6 USN, CO, VF-11, as senior squadron commander remaining in FORRESTAL, became acting COMCVW-17.

221. That CDR B-6 instructed LCDR B-6 , Air Wing Operati Officer, to brief CDR DERRICK on any items of unfinished business on continuing business which might come up during his, B-6 s, abs

222. That on 29 June 1967, the Weapons Coordination Board, augment by key members of the Weapons Planning Board, met to document those ordnance handling procedures which had been validated during the OR

223. That several attendees at this meeting were not members of ei board, but were Warrant Officers and Chief Petty Officers from seve of the squadrons and representatives from ordnance handling divisi of the ship, who were invited as experienced individuals.

224. That LT B-6 attended the 29 June meeting as the repre of COMCVW-17.

225. That the following personnel from VF-11 attended the 29 June
    LTJG B-6 · Weapons Branch Officer
    AOC B-6 - Aviation Ordnance Supervisor

226. That CDR B-6 USN, CVA-59 Operations Officer, who the 29 June meeting, had the minutes read to all attendees at the conclusion of the meeting to ensure there would be no doubt in an mind as to the procedures covered.

227. That the minutes of the meeting were given to all squadrons divisions in a double-spaced rough draft for their correction of ambiguous statements. Any changes of substance would have to be at a subsequent meeting.

228. That after the double-space rough drafts were returned to the CVA-59 Operations Officer, a draft of the procedures in effect was prepared as a memorandum from the Operations Officer to the Commanding Officer dated 8 July 1967.

229. That the draft memorandum dated 8 July was not disseminated outside the Operations Department of the ship and was not seen by the ship's Commanding Officer.

230. That the aforementioned double-spaced rough and the draft memorandum read, in paragraph 1d:

"d. Plug-in of Rocket Pack Pigtails

(1) Required Procedures (Ref: OP-2210) - "Do not arm launcher armament until just prior to takeoff." On carriers this has been traditionally interpreted as "on the catapult".

(2) Effects of Following Required Procedures - Since the development of the TER rack, and its use for carrying LAU-10/LAU-3 series rocket packs, it is commonplace on Yankee Station for a single aircraft to launch with six or more rocket packs. If the ordnanceman is forced to wait until the aircraft is on the catapult before plugging in the pigtail, the launch rate will inevitably decrease. In these frequent cases where a connection proves to be difficult and requires several attempts to accomplish, this delay will become prohibitively long.

(3) Recommended Deviations - Allow ordnance personnel to connect pigtails "in the pack", prior to taxi, leaving only safety pin removal on the cat.

(4) Additional Safety Precautions - Strictest adherence to existing safety precautions provide the very minimum possibility of inadvertent firing. Ordnancemen must receive positive identification of awareness of pilot (and crewman, where applicable) that connection is being made, and "hands off" signal given and acknowledged. Stray voltage check must be scrupulously performed prior to connecting pigtail. Safety pin will not be removed prior to aircraft being positioned on the catapult, nor

in the case of the LAU-10 will the arm lever be cocked.

(5) Final Recommendation - Allow connection of rocket pack pigtail prior to aircraft taxi to catapult."

231. That LTJG   gave the VF-11 copy of the double-spaced rough to AOC   He told   to insert a change which he,   . had proposed and to return it to LI   for further transmittal to the Ship's Operations Officer.

232. That neither LTJG   or any other officer senior to him in VF-11 saw the changes prepared by

233. That no one in VF-11 kept a copy of the double-spaced rough dated 29 June 1967.

234. That the CO VF-11, CDR   did not see any copy of the 29 June double-spaced rough.

235. That AOC   recommendation on the VF-11 copy of the double-spaced rough draft of the minutes of the 29 June meeting was that non-propulsive units be installed on AIM-9 missiles loaded on aircraft on the hangar deck.

236. That the recommendation of AOC   regarding the use of the non-propulsive unit on AIM-9 missiles installed on aircraft on the hangar deck was incorporated in the 8 July 1967 draft memorandum.

237. That CDR   regarded the apparent reversal not as a change of the procedures agreed to at the 29 June meeting, but as a typist's error made on the 29 June double-spaced rough.

238. That a pencilled notation in paragraph 1 of the 8 July 1967 draft memorandum states: "All members took notes for review and a second meeting was held (blank) for final formulation of recommended procedures". The meeting designated by "(blank)" was not held.

All redactions are B-6

66

239. That LT _____ considered that the procedures in the rough draft of the minutes of the 29 June meeting were in effect, although he believed that a smooth document describing the procedures would be promulgated.

240. That LTJG ____ AOC believed that a final smooth copy, approved by the Commanding Officer, CVA-59, was necessary to formalize the procedures covered at the 29 June meeting.

241. That AOC _____ inquired of LTJG and AOC Air Wing Ordnance CPO) several times before 29 July 1967 whether a finalized approved version of the decisions reached at the 29 June meeting had been received.

242. That LTJG briefed the officers of VF-11, including CDR as to the results of the 29 June meeting during a squadron conference, held for other reasons, just after the completion of the 29 June Weapons Coordination Board meeting.

243. That during July 1967 a number of pilots of VF-11 complained to AO1 VF-11 safety ordnance petty officer, that there were too many personnel around their aircraft during turn-up, and questioned the necessity for a safety ordnance petty officer when ordnance was being connected in the pack.

244. That LTJG and AOC greed that the safety petty officer was not necessary in the pack as long as the plane captains were instructed not to give the pilots any signals to move any switches actuating flaps, hook, etc., actuation of which could possibly cause injuries to the ordnancemen under the aircraft.

245. That CDR nsidered that the purpose of the "hands-off" signal to the pilot, while ordnancemen were working under the aircraft, was (a) to prevent injury to the ordnancemen and (b) to prevent blame of the pilot if any ordnance was accidently fired or dropped.

All redactions are B-6

67

246. That CDR ☼-ʟ verbally approved the policy of eliminating the use of the "hands-off" signal and the ordnance safety petty officer in the pack.

247. That the elimination of the "hands-off" signal in the pack was in contradiction to paragraph 1d(4) of the draft memorandum dated 8 July 1967.

248. That CDR ☼-ʟ and LT ☼-ʟ were not informed of the change in procedure by VF-11 with regard to elimination of the use of the ordnance safety petty officer in the pack.

249. That LT ☼-ʟ did not detect the fact that VF-11 was no longer using the "hands-off" signal or an ordnance safety petty officer in the pack.

250. That although CDR ☼-ʟ considered that it was policy in VF-11 to conduct stray voltage checks after both engines had been started, several VF-11 ordnancemen ☼— ☼ believed that the stray voltage test could be made anytime after power was applied to the aircraft.

251. That specific instructions as to precisely when, how and why the rocket circuitry of an F-4 aircraft should be tested for stray voltage was not contained in any publication available on board FORRESTAL.

252. That VF-11 ordnance crews were not instructed to manually check during pre-flight inspection that the LAU-10/A shorting device was fully forward in the "safe" position, since a visual check of its position was believed to be satisfactory.

253. That ☼-ʟ on occasions, had checked TER safety pins by pulling the pin outward about one-half inch to test the catch. He had also checked pins by releasing the button to determine if the pin could be drawn all the way out.

254. That VF-11 commenced transition from Model F-8 aircraft to Model F-4 aircraft in July 1966.

68

255. That VF-11 changed its homeport from NAS, Cecil Field, Florida to NAS, Oceana, Virginia Beach, Virginia on 15 July 1966.

256. That the last effective organization manual for VF-11 was that contained in COMFAIRJAX Instruction P5440.8A of 29 October 1964, Subject: Standard Squadron Organization.

257. That VF-11 was in the process of writing a new organization manual and squadron instructions, which had not been issued on 29 July.

258. That there were an adequate number of personnel assigned to the Weapons Branch, Aircraft Maintenance Department, VF-11 (Assigned - 26; Allowance - 31) to properly perform the duties of the Branch.

259. That during operations on Yankee Station some TER electrical safety pins were found adrift on the flight deck after launches.

260. That some F-4 aircraft carrying rocket launchers (LAU-10/A or LAU-3A/A) on TER-7, arrived at the catapults without TER electrical safety pins installed.

261. That TER electrical safety pins occasionally failed mechanically and were pulled from the TER racks by the action of the wind over the deck and jet blast on the red warning flags attached to the end of the pins.

262. That it was VF-11 policy to leave the TER electrical safety pin in the TER until the aircraft arrived at the catapult.

263. That two squadrons of CVW-17 (VA-65 and VA-106) were in the practice of removing TER electrical safety pins aft in the pack when they were carrying bombs on the TER's.

264. That the above mentioned squadrons did not interpret the restrictions in the 8 July memo on removing TER electrical safety pins in the pack as applying to TER's loaded with bombs.

265. That CDR ⬚ ⬚ ⬚serve the right to fly with all of his squadrons, but directed that each of the pilots on his staff fly with a specified squadron.

266. That LCDR ⬚ ⬚ was directed to fly with VF-11.

267. That LCDR ⬚ ⬚ had extensive experience in the F-4 model aircraft; most recently as an instructor in VF-101 (Replacement Air Wing Fighter Squadron 101).

268. That LCDR ⬚ ⬚ last took an F-4 NATOPS check on 18 November 1966.

269. That LCDR ⬚ ⬚ attended some, but not all, of the training sessions conducted by VF-11 during the transit to WESTPAC on use of various types of ordnance.

270. That LCDR ⬚ ⬚ had never carried or fired ZUNI rockets until assigned to do so on 29 July 1967.

271. That during the five days that FORRESTAL conducted combat operations on Yankee Station, 486 aircraft were launched; 379 were loaded with ordnance; and only 3 VF-11 aircraft had carried rockets.

FINDINGS OF FACT

SECTION VI

FACTS RELATING TO ORDNANCE AND ARMAMENT PERTINENT TO THE F-4B AIRCRAFT

272. That F-4B #110 assigned to VF-11 was configured for the scheduled 1100H launch on 29 July 1967 as listed below:

External Stores Station	Armament Equipment	Items Carried
No. 1 (Port Outboard)	Outboard Wing Pylon	Empty
No. 2 (Port Inboard)	LAU-17/A Pylon LAU-7/A Missile Launcher with adapters (Port side) LAU-17A Pylon Adapter A/A 37B-5 Triple Ejector Rack (P/N 5821520-527) 3 LAU-10/A Launchers	Items Below 1 AIM-9B 1 TER-7 3 LAU-10/A 4 ZUNI each with M414A1 VT nose fuze and M191 MOD 1 base fuze, 1 Frangible Fairing each
No. 3 (Port Aft Fuselage)	Aero 7A Launcher	1 AIM-7E
No. 4 (Port Forward Fuselage)	Aero 7A Launcher	Empty
No. 5 (Centerline)	Aero 27A Bomb Rack	1 600 gallon External Fuel Tank (full)
No. 6 (Starboard Forward Fuselage)	Aero 7A Launcher	1 AIM-7E
No. 7 (Starboard Aft Fuselage)	Aero 7A Launcher	Empty
No. 8 (Starboard Inboard)	LAU-17/A Pylon LAU-7/A Missile Launcher with Adapters (Stbd side) LAU-17/A Pylon Adapter A/A 37B-5 Triple Ejector Rack (P/N 5821520-527) LAU-10/A Launcher	Items Below 1 AIM-9B 1 TER-7 3 LAU-10/A 4 ZUNI each with M414A1 VT nose fuze and M191 MOD 1 base fuze, 1 Frangible Fairing each
No. 9 (Starboard Outboard)	Outboard Wing Pylon	Empty
Total Ordnance Items		2 AIM-9B Sidewinder 2 AIM-7E Sparrow III 24 ZUNI

273. That Section X of Conventional Weapons Loading Manual, F-4B, NAVWEPS 01-245FDB-75, dated 1 October 1965 (latest change 9 September 1966), prescribed detailed procedures for the loading and arming of rockets on the F-4B aircraft.

274. That the latest issue of NAVAIR 01-245FDB-75 dated 1 July 1967 was not received aboard FORRESTAL until 5 September 1967.

275. That Conventional Weapons Loading Checklist, F-4B, NAVWEPS 01-245FDB-75-7, dated 1 October 1965 (latest change 1 April 1966), provided a checklist for the loading and arming of rockets on F-4B aircraft.

276. That procedures prescribed for stray voltage checks on F-4 inboard stations are contained on pages 10-7 and 10-8 of NAVWEPS 01-245FDB-75 (enclosure (364)) and pages 10-12, NAVWEPS 01-245FDB-75-7 (enclosure 363)). These checks are performed on the LAU-17/A pylon, and test the Sparrow III rocket motor firing circuit and the wing pylon explosive bolt circuit only.

277. That neither a requirement nor a procedure which tests for stray voltage of rocket firing circuits in inboard stations is contained in enclosures (363) or (364).

278. That NAVWEPS OP 2210 Volume 1, Aircraft Rockets, requires that a stray voltage test, following instructions for such tests, be performed prior to connecting the rocket harness to the launcher (Sub-para 7-6.5.4, enclosure (365)). It also states, as a general precaution for aircraft rocket launcher packages, an injunction to never connect a launcher to the aircraft without first making a stray voltage check: page xviii, enclosure (366).

279. That VF-11 had devised a procedure and fabricated a test harness, prior to 29 July, for the purpose of conducting a stray voltage check of the rocket firing circuits of the F-4B/TER combination.

280. That the procedure actually followed by VF-11 ordnance personnel consisted of connecting a Missile Launcher Stray Voltage Test Set to each of the rocket harnesses of the TER, at any time when electrical power was available to the TER, as indicated by illumination of the HOMING INDICATOR LIGHT, and testing each station in turn for the presence of stray voltage prior to connecting the rocket harness into the LAU-10/A receptacle.

281. That since the procedure described above was conducted with the TER electrical safety pin "IN" and the weapons select switch on "BULLPUP", it tested for the presence of stray voltage only in the rocket harness of the TER-7 and in a short wire leading to the select rockets relay switch in the TER-7.

282. That the above described procedure for testing the rocket circuitry for stray voltage did not violate any written instructions. However, NAVAIRSYSCOMHQ messages 052222Z AUG 67 and 091903Z AUG 67 (enclosure (367), received after the 29 July accident) recommend that stray voltage checks be conducted after engines were started, when aircraft was operating on internal power, rather than at any time that electrical power was available to the TER.

283. That NAVAIR 11-75A-40, Technical Manual, TER-7, dated 15 August 1966, was inaccurate in that it did not show the actual wiring of the safety switch in the -527 TERs in use aboard FORRESTAL. Change to NAVAIR 11-75A-40 dated 15 April 1967, which shows the wiring correctly, was not received aboard FORRESTAL until after 29 July 1967.

284. That the change in wiring of the safety switch in the -527 TER (wiring the safety switch to ground rather than in series as in the -505 and -521 TERs) made it possible for a high magnitude current to damage or burn out the safety switch.

73

285. That the direction of throw of the HOME-STEP switch on the -527

TER is reversed from the direction of actuation on the -505 and -521

TERs. See photographs, enclosure (368).

286. That tests of the LAU-10/A shorting device, as described in

FORRESTAL message 041015Z AUG 67 and ORISKANY message 060340Z AUG 67

(enclosure (369)) demonstrated that the shorting device, described in

enclosure (371), pages 3-1 and 3-2 of NAVWEPS OP 2210 Volume 2, ZUNI

AIRCRAFT ROCKET, is not reliable.

287. That the piece of metal taken from the right chest of ADJ3 KNIGHT

was identified by the Naval Ordnance Laboratory (NOL), White Oak,

Maryland, as being a part of a M414A1 VT nose fuze, as used in the

MK24 warhead of a ZUNI rocket. Details are contained in NOL messages

221536Z AUG and 011603Z SEP 1967 (enclosure (373)) and enclosure (406).

288. That cook-off tests of ordnance involved in the FORRESTAL fire,

requested by FORRESTAL 161048Z and 170350Z AUG 67 (enclosure (374))

confirmed that the cook-off time for the AN-M65A1 bomb is on the order

of 85 to 120 seconds. This time is significantly less than for other

bombs tested (see NAVAIRSYSCOMHQ 011710Z SEP 67, enclosure (375)) and

enclosure (405).

289. That one CBU electrical cable, which is physically interchangeable

with the LAU-10/A rocket harness, found on FORRESTAL after 29 July 1967,

was not coded for color or feel identification. Use of a CBU cable

instead of a rocket harness on a LAU-10/A launcher could lead to

inadvertent firing of rockets (see NAVAIRSYSCOMHQ 052222Z AUG 67,

enclosure (367), FORRESTAL 130945Z AUG 67, enclosure (376), and

NAVAIRSYSCOMHQ 251436Z AUG 67, enclosure (377)).

290. That analysis of the kinescope of the PLAT recording of the

sequence of events on 29 July demonstrated that the white puff or

flash which seemed to be off the port bow at 10-51-21H was caused by

an image which appeared to be between the camera lens and the forward

flight deck.

291. That experiments were conducted on August 15, 1967 wherein the PLAT camera was trained in the same direction that it had been at 10-51-21H on 29 July, and flash bulbs were fired from locations on the port side of the flight deck aft. These experiments demonstrated that the PLAT camera recorded a reflection from the port side of the flight deck aft on the plexiglass enclosure of the PLAT booth. The demonstration proved conclusively to the Board that images seen in the film were reflections of the last portion of the flight of the ZUNI rocket as it traveled across the flight deck and struck A-4 #405 (see enclosure (138)).

292. That although the ZUNI rocket in the LAU-10/A launcher is certified as HERO safe in enclosure (378), NAVWEPS 16-1-529 of 15 April 1966, there is a possibility that it may be susceptible to RADHAZ if:

a. The shorting device is defective, and there is no contact between the LAU-10/A contact screw and the rocket motor contact band.

b. A previously unconsidered source of RF energy such as a transmitting ALQ-51 in a nearby aircraft is taken into account. FORRESTAL message 201155Z AUG 67 to NAVAIRSYSCOMHQ requested that these conditions be tested. (Enclosure (379)).

293. That three different ordnance safety pins in service in the Navy; namely, the Aero 7 Sparrow launcher pin, the LAU-17/A pylon adapter pin, and the TER electrical safety pin, will fit the TER rack; however, only the latter will actuate the safety switch reliably (see enclosure (380)), NAVWEPCEN CHINA LAKE 112101Z AUG 67 and paragraph 3, enclosure (377) which is NAVAIRSYSCOMHQ 251436Z AUG 67.

294. That enclosure (381), page 12 of NAVWEPS OP3347, USN Ordnance Safety Precautions, contains statements or requirements which are vague, incomplete, or inapplicable to modern attack carrier aircraft. For example, Section 3, Aircraft Launchers, of Chapter 2 states:

a. (Sub-paragraph 3 a(1))."Loading crews shall make positive check that the aircraft battery and armament switches are in the OFF position prior to loading." Comment: Few attack carrier aircraft have batteries.

b. (Sub-paragraph 3 a(2)). "All external power to the aircraft shall be removed and stray voltage check made prior to loading." Comment: The statement does not specify either the circuit(s) to be tested or the purpose of the check. The omissions are significant because rocket launchers such as the LAU-10/A have generated a need for assuring that stray voltage does not exist in either the rocket firing circuit or the bomb release circuit. Further, interpreted literally, the stray voltage check is to be made after external power to the aircraft is removed. In the case of A-4, F-4, and F-8 aircraft, since engines are not running during loading, one would not expect to detect stray voltage since those aircraft would then have no electrical power.

c. (Sub-paragraph 3d). "The pigtail shall not be plugged into the launcher receptacles until just before take-off." Comment: The phrase "just before take-off" is vague both as to time and place. Further, the statement does not contain a requirement for completing a stray voltage check of the rocket firing circuit prior to connecting the rocket harness to the launcher receptacle.

FINDINGS OF FACT

SECTION VII

TRAINING PROCEDURES AND MATERIAL CONDITIONS AS RELATED TO

FIRE FIGHTING AND DAMAGE CONTROL

295. That the underway refresher training period for USS FORRESTAL was shortened from the normal 6 weeks for attack carriers to a period of 4 weeks.

296. That during REFTRA FORRESTAL was graded Unsatisfactory in setting material condition Zebra, but because of good progress in all other damage control areas, received an overall mark of Satisfactory.

297. That concentrated training after REFTRA and prior to the ORI resulted in a grade of Satisfactory for setting material condition Zebra for the ORI.

298. That 37% of personnel in the damage control organization who went through refresher training were transferred, and replaced subsequent to REFTRA and prior to arrival Yankee Station.

299. That upon arriving on Yankee Station, 1610 officers and men, members of FORRESTAL ship's company, had attended a fire fighting school of at least two days duration within the past 36 months. This represents 57% of the ship's company.

300. That during the period from overhaul to Yankee Station, FORRESTAL was at sea 106 days and exercised at general quarters 57 times for an average of one exercise every 1.8 days.

301. That the status of material readiness on 29 July 1967 as relates to damage control and fire fighting was:

    a. Two electric fire pumps down for maintenance.

    b. HCFF Station 5 out of commission in the automatic mode, but operable manually.

    c. Three electrical submersible pumps inoperative.

302. That upon arrival at REFTRA 17 February 1967, FORRESTAL HCFF systems were completely unreliable, however, they were at that time taken apart one-by-one, completely disassembled, restoring all but station 5 to good operating condition. HCFF station 5 is still inoperative in automatic mode due to a faulty HYTROL valve, but continues to be operative manually.

303. That FORRESTAL was up to allowance in essential damage control equipment but had minor deficiencies in repair lockers such as wedges, spanner wrenches, and plugs.

304. That the on-hand inventory of selected pertinent damage control equipment just before the fire was as follows:

	ALLOWANCE	ON HAND BEFORE FIRE
OBA's	550	525
OBA CANNISTERS	3300	3100
1½" NOZZLES	278	278
1½" HOSE	646 (50' length)	646
2½" NOZZLES	52	52
2½" HOSE	105 (50' length)	105
FOG FOAM	1220 CANS	1170 CANS

305. A large number of OBA's, OBA cannisters, and cans of fog foam were received from ships in the vicinity during the fire. Exact quantities unknown.

306. That all damage control discrepancies from the INSURV inspection were corrected with the exception of a shortage of MK 5 gas masks, which shortage continues to exist.

307. That the status of HCFF hoses on the flight deck at the time of the first major explosion at 10-52-55H was:

Station	Frame #	Hose Length	Status
1 stbd	28	100'	Faked, uncharged, operable
1 port	24	100'	Faked, uncharged, operable
2	61	200'	Faked, uncharged, operable
3	---	---	No fitting on flight deck
4	100	150'	Faked, uncharged, operable
5	73	250'	Out of commission in "automatic"
6	118	200'	Faked, uncharged, operable
7	113	250'	Being led aft, uncharged, operable
8	146	150'	Being led aft, uncharged, operable
9	137	300'	Led aft 50' short of fire, charged
10	162	100'	Led to vicinity #316, charged
11	167	200'	Being led across deck, actuated but not charged
12	181	150'	Being led aft, uncharged
13	176	100'.	Being led across deck, uncharged
14	204	150'	Engulfed in flames
15	226	100'	Being led across deck, uncharged
16	---	---	No fitting on flight deck
17	230	100'	Faked, uncharged, operable

308. That the status of salt water fire hoses on the flight deck at the time of the first major explosion was:

Station	Frame #	Hose Length	Status
1	5	100'	Faked, uncharged, operable
2	5	100'	Faked, uncharged, operable
3	32	100'	Faked, uncharged, operable
4	32	100'	Faked, uncharged, operable
5	62	200'	Faked, uncharged, operable
6	53	200'	Faked, uncharged, operable
7	81	100'	Faked, uncharged, operable
8	104	150'	Faked, uncharged, operable
9	119	300'	Faked, uncharged, operable
10	127	150'	Being led aft, uncharged, operable
11	145	250'	Being led aft, uncharged, operable
12	137	150'	Being led aft, uncharged, operable
13	174	250'	Being led aft, uncharged, operable
14	154	100'	Led out vicinity #316, charged
15	212	150'	Led out vicinity #416, charged, and playing on fire
16	186	150'	Led out vicinity of #416, uncharged
17	227	100'	Being led out across deck, uncharged, operable
18	214	100'	Engulfed in flames
19	----	----	No fitting on flight deck
20	226	100'	Engulfed in flames

309. That at the start of the fire, four of the installed sixteen HCFF stations were manned on the second deck. These were stations 2, 7, 9, and 14. These are the stations which are normally manned during flight quarters.

310. That at the commencement of the fire, word was immediately passed over the 1MC "man all fog foam stations."

311. That numerous personnel who were near the fire fighting stations at the outset of the fire on the flight deck were unfamiliar with fire fighting procedures and therefore unable to contribute to the fire fighting efforts.

312. That many personnel did not reach their general quarters stations. Some personnel made no attempt to proceed thereto while others were unable to because of injury; because the effects of the casualty prevented their movement, or, because they became effectively engaged in other required tasks, such as fire fighting.

313. That difficulty was experienced in promptly leading out and charging HCFF stations 11, 12 and 13 in the early stages of the fire.

314. That personnel experienced difficulty in leading the hose from HCFF station 11 toward the fire on the flight deck because the hose became entangled in yellow equipment on deck in the vicinity of the RA-5Cs; starboard side, abaft the island.

315. That, although the valve was open, and the actuation button pressed, the hose at HCFF station 11 was not charged by the time the first bomb exploded.

316. That one or more fog foam stations were not initially charged because personnel manning the hose either did not know it was necessary to perform certain manual functions on the flight deck to actuate the system, or, did not know how to do so.

317. That the HCFF hose stations vary considerably in configuration throughout FORRESTAL. At some stations the sound powered handsets are in boxes; at others the switches are in boxes. The physical relationship of the valve, actuating button, phone, and call button follows no established pattern. Frequently there are similar fittings not associated with the HCFF system in such close proximity that they can cause confusion.

81

318. That flight deck HCFF hose station 14 and salt water fire hose stations 18 and 20 could not be manned at the outset of the fire because they were immediately engulfed in flames.

319. That flight deck HCFF hose stations 15 and 17 and salt water hose station 17 were not led out or charged at the time of the first major explosion. These hoses were engulfed in flames by the second major explosion.

320. That as the second deck HCFF stations are presently configured, a man on the second deck manning a HCFF station is unable to initiate communications with the crew on either the hangar or the flight deck HCFF stations which are served by his equipment. He has no call button.

321. That on 7 August 1967, a test of HCFF station 2 in USS FORRESTAL was conducted by ship's personnel at the request of the Board of Investigation in order to familiarize members of the Board with the operation of a HCFF station and to determine an order of magnitude of time required to produce fog foam. HCFF Station 2 was randomly selected. HCFF Station 2 performed in an acceptable manner when energized from the remote control switch at the flight deck level. The riser was initially empty. Water was produced at the nozzle in 8 seconds. Foam appeared at the nozzle after a total elapsed time of 17 seconds from activation of the system.

322. That on 22 August 1967, the Board of Investigation tested, without warning, ten HCFF stations on the flight deck for automatic operation. The times required to generate good foam at 8 stations at the flight deck level varied from 30 to 45 seconds. One station failed to generate, producing salt water only. One station required 4 minutes to generate foam.

323. That of the 525 OBAs on hand in FORRESTAL just before the fire, 450 were new, the remainder were of late design and in good material condition.

324. That difficulty was experienced by some personnel in pulling the tabs to open OBA cannisters.

325. That some OBA cannisters did not last the rated 30 minutes.

326. That significant numbers of air wing personnel were not properly checked-out on the use of OBAs.

327. That complaints were received from users of OBAs that some were in poor material condition, i. e., broken straps, holes in the mask and timers not working.

328. That the damage control message chits, which would have provided an invaluable record of events related to the casualty, were inadvertently destroyed, greatly hampering the Board's ability to reconstruct a true picture of the damage control and fire fighting efforts.

329. That only two men were killed during the fire fighting operations after major explosions had subsided. They were probably killed by chlorine gas. ᛒᛔ died in compartment 03-231-0-L after having removed his OBA because the compartment was relatively free of smoke, and ᛒᛔ died in compartment 03-226-0-L at 0300, 30 July 1967 while removing bodies.

330. That during the afternoon of 29 July, inexperienced air wing personnel actuated the divisional doors between hangar bays 2 and 3 (opening and then closing same) without authority of Damage Control Central.

331. That it required one man approximately one hour to jettison 750 gallons of liquid oxygen through a single one inch, 16 foot length hose, at its storage location, compartment 1-192-2-E, near hangar bay 3.

332. That there is no emergency dump capability installed in FORRESTAL's liquid oxygen systems.

333. That the 1MC ship's announcing system was ineffective in the hangar deck areas; thus denying personnel in those areas important information and directives during the emergency.

334. That prior to the fire, FORRESTAL's planning and drill for flight deck fire fighting had concentrated chiefly on the landing areas of the flight deck, with lesser emphasis elsewhere.

335. That only one minor electrical casualty occurred during the entire period of the emergency; the temporary loss of air conditioning in Main Comm and supra 07 radio spaces. In spite of the large number of electrical cables burned by fires, FORRESTAL experienced no secondary electrical fires therefrom.

336. That the escalator between 2nd and 02 decks (2-187-2-L), acted as a reverse stack feeding smoke downward to the second deck.

337. That the identity of damage control and repair party personnel was not evident to large numbers of unengaged personnel in the ship, particularly throughout the hangar bays. Many sought to assist in fire fighting but were unable to identify the on-scene leader.

338. That no significant instability resulted to the ship from excess water in the hull incident to fire fighting.

339. That a minor list of 2.5 degrees to port was caused by shrapnel holes in the transom and on the port side aft near and at the waterline which flooded voids 8-244-2-V, 8-244-4-V, and 8-244-6-V. This list was corrected by transfer of NSFO from port to starboard tanks.

340. A preliminary decision was made to flood magazines 7-222-1-M, 7-222-2-M, 7-227-3-M and 7-231-0-M in the after portion of the hull. Damage Control Central subsequently determined from Air Ordnance Control Station (AOCS) that the ordnance in these magazines was in fact inert. They were not flooded.

341. That the location of MK 24 paraflares on 29 July 1967 at 1051H
and their subsequent disposition is as follows:

Location	Number of Flares	Disposition
4-172-0-M Deep storage magazine	459	Retained
02-146-3-M Taping and Banding Room	24	*Jettisoned
Jettisonable Lockers	96	*Jettisoned
	Total 579	

* 120 MK24 paraflares jettisoned at commencement of the fire; no flares
actually involved in the fire.

342. That during the fire, normal services were maintained in the un-
damaged areas of FORRESTAL including light, ventilation, air conditioning,
hot and cold water.

343. That there is no known survival training curriculum directed toward
training aircrewmen in abandoning static aircraft engulfed in flames.

FINDINGS OF FACT

SECTION VIII

FATALITIES AND INJURIES

344. That 134 persons, 116 whose remains have been positively identified,
are dead or missing as the result of the fire and explosions in FORRESTAL
on 29 July 1967. The names, location, cause of death and remarks concerning the cause of death of all fatalities are contained herein at the end
of this section.

345. That the 18 persons who are missing or whose remains are unidenti-
fiable and are presumed dead (see enclosure (362)) are:

PR2, USN

USN

AEAN, USN

DJ3, USN

, USN

USN

USN

I, USN

ORAN, USN

AE3, USN

IC, USN

AMM2, USN

ATN3, USN

USN

AMS2, USN

SN, USNR

, LCDR, USN

AN, USN

All redactions
are B-6

346. That 12 to 14 remains were recovered aboard FORRESTAL, but were not
identifiable.

690105-0011

347. That 4 persons have been determined missing in the water without recovery of remains.

348. That 3 of these persons,   .   .   were seen jumping into the water from the port quarter at about 1145H.

349. That the fourth man believed lost over the side was sighted jumping from FORRESTAL by the COD, USS RUPERTUS before the initial major explosion.

350. That at least 16 other FORRESTAL personnel were blasted, knocked, jumped or fell into the water on 29 July 1967 and were subsequently recovered. Thirteen of these have been identified as follows:

NAME	TIME IN WATER	TIME OUT OF WATER	METHOD OF RECOVERY	DISPOSITION
	1130	1200	Motor Whale Boat	MACKENZIE ORISKANY CUBI POINT FORRESTAL
	1200	1230	Helo	ORISKANY CUBI POINT FORRESTAL
1.	1135	1155	Helo	ORISKANY FORRESTAL
	1115	1145	Helo (#2)	FORRESTAL
	1150	1205	Helo	ORISKANY CUBI POINT FORRESTAL
	1140	1155	Motor Whale Boat	MACKENZIE ORISKANY CUBI POINT FORRESTAL
	1145	1205	Motor Whale Boat	MACKENZIE ORISKANY CUBI POINT FORRESTAL
	1145	1205	Motor Whale Boat	MACKENZIE ORISKANY CUBI POINT FORRESTAL
	1100	1115	Motor Whale Boat	MACKENZIE ORISKANY CUBI POINT FORRESTAL

All redactions are B-6.

87

NAME	TIME IN WATER	TIME OUT OF WATER	METHOD OF RECOVERY	DISPOSITION
	1130	1145	Motor Whale Boat	MACKENZIE ORISKANY CUBI POINT FORRESTAL
	1100	Unknown	Helo	ORISKANY REPOSE EXPIRED
	1100	Unknown	Unknown	OKALAND ARMY BASE MORTUARY EXPIRED
	Approx. 1100	Approx. 1115	Helo	ORISKANY REPOSE MED-EVAC

351. That   amped into the water with three other men, identities unknown, and all were picked up and taken to USS ORISKANY by helicopter.

352. That a total of 27 personnel were killed or injured by the first bomb explosion while engaged in fire fighting efforts on the flight deck. Nine of these personnel were killed, 4 of whom were V-1 crash crew members (see enclosure (316)). Sixteen personnel were injured, 5 of whom were V-1 crash crew members.

353. That V-1 crash crew personnel,   nd were killed by the first bomb explosion while fire fighting.

354. That V-1 crash crew personnel   and were injured by the first bomb explosion while fire fighting. and   later died.

355. That non-crash crew personnel,   V-1;   V-2;   IA-65; VA-65; and   V-1, were killed by the first bomb explosion while fire fighting.

All redactions are B-6.

356. That V-1 Division non-crash crew personnel,

.   .   .   and   . were injured by

the first bomb explosion while fire fighting.

357. That other non-crash crew personnel,   ___ , VF-11;   , VF-11;

___ R, VA-106;   ,   1; and   RVAH-11 were injured by

the first bomb explosion while fire fighting.

358. That a total of 161 personnel were injured as a result of the fire

and explosions (see enclosures (328) and (329)).

359. That 87 of the personnel injured were hospitalized for a period of

24 hours or longer for injuries received as a result of the fire and

explosions (see enclosure (329)).

360. That 74 of the personnel injured were treated for injuries received

as a result of the fire and explosions and returned to duty without

hospitalization (see enclosure (328)).

All redactions are B-6.

NAME	LOCATION/CAUSE OF DEATH	REMARKS/DOCUMENTED BY ENCL NR
	AME Shop (Hangar Bay 3) Presumed drowned.	First seen in AME shop, then made way to port quarter where he went overboard with raft. (220) to (225), (362).
Division	2-244-2-L Multiple extreme injuries.	Presumed trapped in compartment. (226), (362).
VF-11	03-226-0-L Multiple extreme injuries.	Presumed trapped in compartment. (216), (362).
?-.74	1 03-241-1-L Multiple extreme injuries	Presumed trapped in compartment. (227), (362).
VA-46	Flight deck. Cause of death unknown	Escaped from aircraft and was running forward when first bomb exploded. (228), (362).
MH2 F-11	03-231-0-L Multiple extreme injuries	Presumed trapped in compartment. Encl (216), (362).
VF-11	03-226-0-L Third degree burns over 80% of body.	Presumed trapped in compartment. (216), (362).
SN, Div.	3-237-2-E Multiple extreme injuries.	Standing watch in port after steering, was wounded by shrapnel during initial explosions. Continued to man his station and sound powered phone till 1130 when contact with him was lost. (229), (362).
4A,	03-236-0-L Burns, 3rd degree over 80% of body.	Presumed trapped in compartment. (217), (362).
3 VF-11	03-226-0-L Third degree burns entire body.	Presumed trapped in compartment. (230), (362).
AMS1 F-11	03-241-2-L Explosion and fire.	Presumed trapped in compartment. (216), (362).
AE3 F-11	03-236-0-L Multiple extreme injuries.	Was awakened and last seen compartment. (231), (362).

All redactions are B-6.

NAME		LOCATION/CAUSE OF DEATH	REMARKS/DOCUMENTED BY ENCL NR
L., AQF3, F-74		03-236-0-L Third degree burns 60% - 70% body.	Presumed trapped in compartment. (232), (362).
., PRAN, F-74		03-236-0-L Multiple extreme injuries.	Presumed trapped in compartment. (217), (362).
. A., SN A Div.		03-207-2 Lung damage due to smoke inhalation.	Had been fire fighting since fire began, helping recover bodies when overcome at approximately 0330, 30 July 1967. (233), (234), (362).
2 VF-11		Flight Deck. Multiple extreme injuries.	Aft in vicinity of A/C 110. (235) to (239), (362).
	, SN, F-74	03-236-0-L Third degree burns 70%-80% body.	Presumed trapped in compartment. (217), (362).
	AN, F-11	AMD Para-loft. Multiple extreme injuries.	Presumed trapped in para-loft. (240), (362).
ABH2, VF-74		03-236-0-L Third degree burns 70%-80% body.	Presumed trapped in compartment. (241), (362).
AME2, VF-74		03-236-0-L Multiple extreme injuries.	Presumed trapped in compartment. (217), (362).
	ABAN, F-74	03-236-0-L Cause of death unknown.	Presumed trapped in compartment. (217), (362).
	AO2, -65	Ordnance Shop (03-236-2) Traumatic amputation both legs, third degree burns 60% body.	When fire broke out he was in Ordnance Shop (03-236-2). Last seen running forward about midships. (242), (362).
	AE3, F-11	03-226-0-L Third degree burns entire body.	Presumed trapped in compartment. (216), (362).
	ADJ3, >-<	03-231-0-L Cause of death unknown.	Presumed trapped in compartment. (219), (362).
	ADJ3, F-11	03-226-0-L Third degree burns entire body.	Presumed trapped in compartment. (216), (362).

All redactions as T-6

NAME	LOCATION/CAUSE OF DEATH	REMARKS/DOCUMENTED BY ENCL NR
ADJ3 VF-74	03-236-0-L Multiple extreme injuries.	Presumed trapped in compartment. (217), (362).
, SN, VF-11	03-226-0-L Third degree burns 70% body.	Presumed trapped in compartment. (216), (362).
, AE1, VF-74	03-241-0-L Cause of death unknown.	Presumed trapped in compartment. (217), (362).
, AO2, VA-106	03-236-1-L Multiple extreme injuries.	Presumed trapped in VA-106 ordnance berthing compartment. (218), (362).
ATN2 , VF-11	AMD Para-loft Multiple extreme injuries.	Presumed trapped in Para-loft. (244), (362).
..., ABHC, V-1 Div.	Flight Deck. Multiple extreme injuries.	Attempting to extinguish fire on flight deck with PKP bottle when 1st explosion occurred. (245), (299), (362).
, FN, : Div.	2-237-2-E Multiple extreme injuries.	Standing watch in port steering. Early explosion cut both arms. First aid was given on the spot by the Quartermaster. (246), (247), (362).
AME1, VF-74	AME Shop (1-217-4) Death by drowning is presumed.	Trapped in AME Shop(1-217-4) till approximately 1215 when worked way to port quarter. Jumped from Port Quarter. (248), (362).
\N, .?-74	03-236-0-L Cause of death unknown.	Presumed trapped in compartment. (217), (362).
..., ..., DJ1, VA-106	Flight Deck. Multiple extreme injuries.	Last seen early stages of fire proceeding forward along starboard catwalk. (249), (362).
AN, S-2	03-231-0-L Cause of death unknown.	Presumed trapped in compartment. (219), (362).
V-1	Flight Deck. Multiple extreme injuries.	Hot suit man. Ran directly to fire and was fighting the fire when 1st explosion occurred. (299), (362).

All redactions are B 6

3

NAME		LOCATION/CAUSE OF DEATH	REMARKS/DOCUMENTED BY ENCL NR
, AMS1, F-74		03-236-0-L Multiple extreme injuries.	Presumed trapped in compartment. (217), (362).
(NMN), AN F-74		03-236-0-L Multiple extreme injuries.	Last seen beside his rack dressing after General Quarters sounded. (250), (362).
., ----, LT, 4-65		Flight Deck. Explosion and fire.	Last seen running toward the fire after the first explosion. (251), (362).
, ADJ3, F-11		03-226-0-L Multiple extreme injuries.	Presumed trapped in compartment. (216), (362).
AMEAN, VF-74		03-(231 or 226)-0-L Third degree burns burns 70%-80% body.	Presumed trapped in compartment. (252), (362).
, AMS3, VF-11		03-231-0-L Third degree burns entire body.	Presumed trapped in compartment. (216), (362).
AMS3 F-74		03-236-0-L Cause of death unknown	Presumed trapped in compartment. (217), (362).
AE3		03-236-0-L Third degree burns entire body.	Presumed trapped in compartment. (253), (362).
., ATR3 VF-74		03-236-0-L Third degree burns entire body.	Presumed trapped in compartment. (254), (362).
SN, Div.		4-244-0-A Multiple extreme injuries.	Remains recovered approximate 1300 30 July 1967. Compartment 4-244-0-A had been flooded, apparent cause of death suffocation or drowning. (255), (256), (362).
-1 Div.		Flight Deck. Multiple extreme injuries.	Flight deck port side aft by cats 3 and 4, just before 1st explosion. (257), (362).
VA-46		Flight Deck. Shrapnel wound abdomen, traumatic amputation left arm third degree burns 50% body.	Arm severed before first explosion, possibly by ZUNI Rocket on its path across flight deck. He was taken to island and later to ORISKANY where he died. (258), (259), (362).

* 6

- 8

(40)

All redactions are B-6.

4

NAME	LOCATION/CAUSE OF DEATH		REMARKS/DOCUMENTED BY ENCL NR
., AN JF-11	AMD Para-loft Multiple extreme injuries.		Presumed trapped in AMD Para-loft. (260), (362).
DS2 OE	03-226-0-L Lung damage due to smoke inhalation.		had been aiding in the effort to remove remains from the 03 level in the early morning the day following the fire when apparently overcome by smoke and/or Chlorine Gas. (261), (362).
, AMH3 VF-11	03-226-0-L Multiple extreme injuries.		Presumed trapped in compartment. (216), (362).
, AE2 -11	AMD Para-loft. Multiple extreme injuries.		Seen standing in his cube, Compartment 03-226-0-L, after the first bomb explosion. Next seen in AMD Para-loft, where he was apparently trapped. (262), (263), (362).
, AN, V-1	Flight Deck. Second and third degree burns 80% of body.		Fighting fire by Salt Water Station #15 when fires forced him down the external ladder near the B & A crane. While going down the ladder he caught fire and _ threw him overboard to extinguish the flames. Was recovered from water by ORISKANY, transferred to REPOSE where he died (264), (265), (362).
., AO3 F-11	Flight Deck. Multiple extreme injuries.		Last seen near aft fuel station just after fire started. (266), (267), (362)
WO1, -1 Div.	Flight Deck. Multiple extreme injuries.		went down the external ladder near the B & A crane with . When caught fire he threw him overboard to put out the flames, then followed him into the water. saw in the water but did not see him recovered by helo. remains were recovered but it is not known how he was picked up out of the water. (264), (268), (362).
, ADJAN, F-74	03-236-0-L Multiple extreme injuries caused by fire and explosion.		Presumed trapped in compartment. (217), (362).

All redactions aff B-6.

91

NAME		LOCATION/CAUSE OF DEATH	REMARKS/DOCUMENTED BY ENCL NR
ATR3, VF-11		03-226-0-L Multiple extreme injuries.	Presumed trapped in compartment. (216), (362).
	AN, 1	Flight Deck. Multiple extreme injuries.	Ran aft when fire broke out to aid firefighting effort. Survived 1st explosion, but was last seen in vicinity of elevator 3 and 4 prior to second explosion. (269), (362).
	TF-74	03-236-0-L Multiple extreme injuries.	Presumed trapped in compartment. (270), (362).
	AN TF-74	03-236-0-L Third degree burns entire body.	Presumed trapped in compartment. (217), (362).
AN	(NMN)III -11	03-226-0-L Third degree burns 70% body.	Presumed trapped in compartment. (271), (362).
	AME3 TF-11	03-226-0-L Multiple extreme injuries caused by fire and explosion.	Presumed trapped in compartment. (216), (362).
	SN, JF-11	03-226-0-L Third degree burns entire body.	Presumed trapped in compartment. (216), (362).
13		Flight Deck. Multiple extreme injuries.	Firefighting on the Flight deck near fog foam station #12. Killed by first explosion and thrown aft of number 3 J.B.D. on Port side. (272), (362).
EM2	E,	2-244-2 Multiple extreme injuries.	Appeared to be heading to port steering space when caught by bomb explosion. (273), (362).
	AN, TF-11	03-226-0-L Multiple extreme injuries.	Presumed trapped in compartment. (274), (362).
	AQF3 TF-11	03-226-0-L Multiple extreme injuries.	Presumed trapped in compartment. (216), (362).
	GFAN, TF-11	03-231-1 Third degree burns entire body, traumatic amputation of extremities.	Presumed trapped in compartment. (216), (362).
	AZ2, TF-11	03-226-0-L Traumatic amputation lower legs, third degree burns 40% body.	Presumed trapped in compartment. (275), (362).

All redactions are B.6.

95

NAME	LOCATION/CAUSE OF DEATH	REMARK/DOCUMENTED BY ENCL NR
., AMH3 .., VF-11	03-226-0-L Multiple extreme injuries.	Presumed trapped in compart- ment. (216), (362).
AQF3, VF-74	03-236-0-L Multiple extreme injuries.	Presumed trapped in compart- ment. (217), (362).
V-1	Flight Deck. Cause of death unknown.	Had finished starting an A-4 aircraft and was sitting on tractor prior to first explosion. The first explosion blew tractor across flight deck and was no longer seen. (276), (277), (362).
?-74 AMH2	03-236-0-L Multiple extreme injuries.	Presumed trapped in compart- ment. (217), (362).
ADJAN, VF-74.	03-236-0-L Multiple extreme injuries.	Presumed trapped in compart- ment. (217), 362).
ADJ1, VF-74	03-241-0-L Third degree burns 70% body.	Presumed trapped in compart- ment. (217), (362).
F-11	03-226-0-L Multiple extreme injuries.	Presumed trapped in compart- ment. (216), (362).
AMH3, -11	03-226-0-L Multiple extreme injuries.	Presumed trapped in compart- ment. (216), (362).
AN, /A-106	03-236-1-L Shrapnel wounds, third degree burns entire body.	Presumed trapped in VA-106 Ordnance Berthing Compartment (218), (362).
AQFAN, F-74	Flight Deck. Multiple extreme injuries.	Last seen on elevator #4 at 1048 29 July 1967. (278), (362).
EAN, -74	03-236-0-L Multiple extreme injuries.	Presumed trapped in compart- ment. (217), (362).
PR3, .., VF-11	AMD Para-loft. Multiple extreme injuries.	Presumed trapped in AMD Para- loft. (279), (362).
AE3 ... /F-11	Flight Deck. Cause of death unknown.	Vicinity of A/C 112 and A/C 110 just prior to fire. (280) to (285), (362).

All redactions are B-6

7

NAME	LOCATION/CAUSE OF DEATH	REMARKS/DOCUMENTED BY ENCL NR
▓ ▓ AE3, ▓	03-226-0-L Third degree burns entire body.	Presumed trapped in compartment. (216), (362).
▓ ▓, AN, F-74	03-236-0-L Third degree burns lower body, traumatic amputation upper 1/3 body.	Presumed trapped in compartment. (217), (362).
▓ AA, F-11	AMD Para-loft. Multiple extreme injuries.	Presumed trapped in AMD Para-loft. (286), (287), (362).
▓ FN, Div.	3-237-2-E Multiple extreme injuries.	Standing watch in port steering. (288), (362).
▓ VA-46	Flight-Deck Multiple extreme injuries.	Last seen near A/C 405 approximately 5 minutes before the fire. (289), (290), (362).
▓ AN, VF-74	03-236-0-L Multiple extreme injuries.	Presumed trapped in compartment. (217), (362).
▓ AMHC, VA-46	Flight Deck. Cause of death unknown	Last seen in vicinity of A/C 410 approximately 5 minutes before the fire. (290), (291), (362).
▓ AMH3, VF-11	03-226-0-L Third degree burns 80% of body.	Presumed trapped in compartment. (216), (362).
▓ ▓02, A-106	03-236-1-L Multiple extreme injuries.	Presumed trapped in VA-106 Ordnance berthing compartment. (218), (362).
▓ PRAN, VA-46	03-177-0-L Multiple extreme injuries.	Seen leaving compartment 03-177-0-L about 5 minutes before fire. (289), (362).
▓ AMH2 F-11	03-241-0-L Cause of death unknown.	Presumed trapped in compartment. (292), (362).
▓ AN F-74	03-236-0-L Third degree burns 70%-80% of body.	Presumed trapped in compartment. (293), (362).
▓ AE3 VF-74	03-226-0-L Third degree burns 70%-80% of body.	Presumed trapped in compartment. (216), (362).

All redactions are B-6

NAME	LOCATION/CAUSE OF DEATH	REMARKS/DOCUMENTED BY ENCL NR
, ATN3, VF-11	03-226-0-L Cause of death unknown.	Presumed trapped in compartment. (216), (362).
, ABH3, VF-74	03-236-0-L Third degree burns entire body.	Last seen running through compartments waking men. (294), (362).
., AOAN, VA-106	03-236-0-L Multiple extreme injuries.	Presumed trapped in VA-106 Ordnance berthing compartment (218), (362).
., AMS3, VF-11	03-226-0-L Multiple extreme injuries.	Presumed trapped in compartment. (216), (362).
AME2, °-11	AME Shop, 1-217-4 Death by drowning is presumed.	First seen in AME Shop, then made way to port quarter where he went overboard with a one-man raft. (220) to (225), (362).
AN, F-11	03-226-0-L Multiple extreme injuries.	Presumed trapped in compartment. (216), (362).
VF-11	03-226-0-L Third degree burns 70%-80% body.	Presumed trapped in compartment. (216), (362).
, AO2, VA-65	ORN Shop 03-236-2. Third degree burns entire body.	When fire broke out he was i: ORN Shop (03-236-2). Last seen running forward about midships. (242), (362).
., AMH3, 06, VF-11	03-226-0-L Cause of death unknown.	Presumed trapped in compartment. (216), (362).
ABH3, -1 Div.	Flight Deck. Multiple extreme injuries.	Last seen going into area of 1st fire, just prior to 1st explosion. (299), (362).
AOAN, F-11	03-226-0-L Multiple extreme injuries, third degree burns 90% body.	Presumed trapped in compartment. (216), (362).
SN,	03-236-0-L Multiple extreme injuries.	Presumed trapped in compartment. (217), (362).
AMS2, VF-74	03-236-0-L Cause of death unknown.	Presumed trapped in compartment. (217), (362).

All redactions are B6.

98

NAME		LOCATION/CAUSE OF DEATH	REMARKS/DOCUMENTED BY ENCL NR
	O1 -74	03-241-0-L Third degree burns 70% of body.	Presumed trapped in compartment. (217), (362).
	SN, Div.	4-244-0-A Cause of death unknown.	Assigned to clean Landing Force Locker (4-244-0-A) the morning of the fire. (256), (362).
	AO3, F-74.	03-236-0-L Multiple extreme injuries schrapnel wounds.	Presumed trapped in compartment. (295), (362).
	CDR, VA-46	Flight Deck. Cause of death unknown.	Last seen in aircraft. Apparently he was unable to free himself before flames covered his aircraft. (296), (297), (298), (362).
	AN, 1 Div.	Flight Deck. Multiple extreme injuries.	Hot suit man. Last seen going into the area of 1st fire. Was fighting fire just prior to 1st explosion. (299), (362).
ADJ2	VA-106	Flight Deck. Second and third degree burns 80% of body.	Severely burned by initial fuel explosion. Transferred to ORISKANY, then to REPOSE where he died. (300), (362).
	AA, -74	03-236-0-L Multiple extreme injuries, third degree burns 70%-80% body.	Presumed trapped in compartment. (217), (362).
	ABHAN, V-1 Div.	Flight Deck. Multiple extreme injuries.	Ran aft when fire broke out apparently to direct aircraft out of area of fire. Believed killed by the 1st explosion. (301), (302), (362).
	ABH2 V-1 Div.	Flight Deck. Multiple extreme injuries.	Was seen near elevator #3 helping an injured man toward the island just prior to 2nd explosion. (303), (362).
	, AN, F-11	03-226-0-L Multiple extreme injuries.	Presumed trapped in compartment. (304), (362).
	YN3, F-74	03-236-0-L Multiple extreme injuries.	Presumed trapped in compartment. (305), (362).

All redactions are B-6.

99

NAME		LOCATION/CAUSE OF DEATH	REMARKS/DOCUMENTED BY ENCL NR
AMH1	VF-74	03-241-0-L Multiple extreme injuries, third degree burns 70%-80% body.	Presumed trapped in compart-ment. (217), (362).
	AE3, -11	03-231-0-L Multiple extreme injuries.	Presumed trapped in compart-ment. (216), (362).
	AN -2	Starboard passageway about frame 200. Multiple extreme injuries.	Last seen departing compartment 03-207-1-L. (306), (362).
	AN VA-106	03-236-1-L Multiple extreme injuries.	Presumed trapped in VA-106 Ordnance berthing compartmen (218), (362).
	ATNAN, VA-46	In the area of 03-231-3-L Multiple extreme injuries.	Last seen departing the compartment after the initial explosion. (307), (362).
25	AN, VA-46	Flight Deck. Cause of death unknown.	Last seen in immediate vicinity of initial fire. (308), (362).
	AA, VF-74	03-236-0-L Multiple extreme injuries.	Presumed trapped in compart-ment. (217), (362).
26	AME3, VA-65	Flight Deck. Multiple extreme injuries.	Last seen holding fire hose aft of island prior to 1st explosion.
	AQF3, F-11	03-226-0-L Multiple extreme injuries.	Presumed trapped in compart-ment. (216), (362).
	ADJ3, VF-11	AMD Para-loft. Multiple extreme injuries.	Presumed trapped in AMD Para-loft (310), (362).
	AA, F-74	03-236-0-L Third degree burns entire body, multiple extreme injuries.	Presumed trapped in compart ment. (311), (362).
27	LCDR, VA-46	Flight Deck. Multiple extreme injuries.	Believed to have exited his aircraft but caught in 1st explosion. (312), (313), (314), (362).
	AMHAN, VF-74	03-236-0-L Multiple extreme injuries.	Presumed trapped in compart ment. (217), (362).
* 28	AN, VA-46	Flight Deck. Second and third degree burns 80% body.	Was near aircraft 416 when fire broke out. Received severe burn and was trans-ferred to USS REPOSE where he died.

All redactions are B-6.

134

FINDINGS OF FACT

SECTION IX

DAMAGE AND LOSS RESULTING FROM THE FIRE

361. That the damage to FORRESTAL from fire, explosions, smoke and water as a result of the fire on 29 July 1967 is related in detail in enclosures (187) through (215a)

362. In general, the damage to the ship included:

a. Major damage to 01, 02, 03 levels and flight deck, aft of frame 212, including major structural members, bulkheads, decks, armored flight deck, and damage to hull of ship abaft hangar bay 3.

b. All gun mounts and fire control equipment damaged beyond repair.

c. Bomb elevators 19, 20, and 21 rendered inoperative.

d. All electrical cables burned on the B and A crane.

e. All piping systems severly damaged above main deck from frame 212 aft.

f. Damage and destruction of ventilation heating and air conditioning system main deck, 2nd and 3rd decks from frame 212 aft.

g. All electrical equipment and wiring burned completely or damaged by water, main deck, 2nd deck and 3rd deck from frame 212 aft.

h. Arresting gear machinery; N2 plant; Number 3 oxygen plant; and port emergency steering unit damaged.

i. AN/SPN 42 rendered inoperative; LSO console destroyed; 22 cables in main mast affecting Tacan, SPN 43, 1 FF, SPN 9, two ARC 27s and ECM severed by shrapnel; and SPN 43 and SPS 30 antenna reflectors damaged by shrapnel.

j. Nine berthing compartments aft completely destroyed.

k. Jet engine repair and stowage facility destroyed.

l. Aft moving stairway rendered inoperative due to salt water damage to electrical equipment.

363. That the cost of voyage repairs accomplished in FORRESTAL at Subic Bay, R. P. after the fire on 29 July 1967 was $45,000

364. That the preliminary estimate by Naval Ship Systems Command for repairing the structural damage to FORRESTAL resulting from the fire and explosions is $16,000,000.

365. That the following aircraft were destroyed as a result of the fire on 29 July 1967:

TYPE	NUMBER	COST EACH	TOTAL COST
F-4B	7	2,749,000	19,243,000
A-4E	11	781,000	8,591,000
RA-5C	3	5,563,000	16,689,000
TOTAL	21		$44,523,000

366. That a total of 40 other aircraft were damaged during the fire on 29 July 1967. Some repairs have already been accomplished by WESTPAC activities (see enclosure (194)), but the total cost of repairs to aircraft is unknown as of 15 September 1967.

367. That the estimated value of ordnance destroyed or jettisoned during the fire on 29 July 1967 is $1,950,000.

368. That the estimated value of supplies and equipage jettisoned, or damaged beyond repair during the fire on 29 July 1967 is $3,150,000.

369. That the estimated cost of repair of equipage damaged during the fire on 29 July 1967 is $225,000.

370. That the replacement cost for the Dual Purpose Battery and associated fire control equipment is $6,210,000.

371.   That as of 15 September 1967 estimates have not been received for the replacement costs of the equipments and systems under the cognizance of the Naval Air Systems Command Headquarters.

372.   That the total of all known estimated costs as of 15 September 1967 is $72,203,000.

373.   That personnel claims for reimbursement for property lost or damaged is unknown.

OPINIONS

1. That the fire on 29 July 1967 aboard FORRESTAL was caused by the accidental firing of one ZUNI rocket from the port TER-7 on external stores station 2 of F-4B #110, VF-11, then spotted on the extreme starboard quarter of the flight deck. The above mentioned ZUNI rocket struck A-4 #405, rupturing its fuel tank, igniting the fuel and initiating the fire.

2. That material failures of aircraft and armament components of F-4B #110 and its loaded ordnance stores, were the proximate cause of the accident.

3. That poor and outdated doctrinal and technical documentation of ordnance and aircraft equipment and procedures, evident at all levels of command, was a contributing cause of the accidental rocket firing.

4. That no improper acts of commission or omission by personnel embarked in FORRESTAL directly contributed to the inadvertent firing of the ZUNI rocket from F-4 #110.

5. That a single ZUNI rocket was fired from the port inboard station of F-4 #110 at 10-51-21H due to the simultaneous malfunctioning of the following components:

    a. CA42282 pylon electrical disconnect.

    b. Safety switch of the TER-7.

    c. LAU-10/A shorting device.

104

6. That, assuming the possible malfunctions described in Opinion 5
above:

a. LCDR ⟨b-b⟩ may have triggered the firing of the ZUNI when
he switched from external to internal electrical power by superimposing
enough transient voltage upon existing stray voltage to fire the ZUNI.

b. ⟨b-b⟩ or ⟨b-b⟩ ay have triggered the firing of the ZUNI
by moving the TER electrical safety pin far enough to arm the TER-7
safety switch while checking the pin to ascertain that it was securely in
place.

c. ⟨b-b⟩ ay have triggered the firing of the ZUNI by mistakenly
"stepping" rather than "homing" the TER-7 after the LAU-10/A launchers
had been plugged in, if a faulty LAU-10/A shorting device were present
on the second or third LAU-10/A in the firing sequence rather than the
first.

7. That the CA42282 pylon electrical disconnect is of defective design
in that it is susceptible to shorting by moisture.

8. That the Naval Air Systems Command erred in:

a. Changing the wiring of the TER-7 mod -527 so that the safety
switch is shorted to ground in the "SAFE" position rather than open
as in models -505 and -521.

b. Issuing the TER-7 mod -527 to the fleet without promulgating
a complete description of the equipment and procedures for conducting
circuit testing and checking based on the new circuitry.

c. Changing direction of throw of the HOME-STEP switch in the
TER-7 mod -527 from that of the TER-7 mod -505 and -521 (see enclosure
(368)).

9. That the electrical safety pin for the TER-7 is poorly designed
in that the ordnance pins which are used in the AERO-7 Sparrow Launcher
and the LAU-17/A pylon, and which will not reliably actuate the TER-7
safety switch, can be mistakenly inserted in the TER-7 pylon.

10. That the existence of ordnance safety pins which are physically interchangeable but not functionally interchangeable creates a potentially dangerous situation.

11. That the LAU-10/A shorting device is of inherently poor design for the following reasons:

    a. There is no positive "safe" position into which the device locks.

    b. This device may be placed in any intermediate position between the extremes of "safe" and "arm" where its actual arm/safe condition cannot be determined by visual inspection. That a very small movement of the slide from "safe" toward "arm" may in fact arm the LAU-10/A, though still appearing to be on "safe". See photographs, enclosure (370).

    c. A bent, short or damaged pin in the 5 pin receptacle of the LAU-10/A may prevent the device from grounding, thereby leaving it armed though appearing to be safe.

    d. That the sliding arm of the device allows interference with the LAU-10/A suspension lug; hence, personnel frequently either cant the device by partially unscrewing it or bend the slide of the device to allow it to clear the lug.

12. That the above inadequacies of the LAU-10/A shorting device constitute an inherently dangerous situation which should be corrected as a matter of urgency.

13. That the transient voltage induced when switching aircraft power of F-4 #110 from "external" to "internal" would not have been sufficient in itself to cause inadvertent firing of the ZUNI rocket. However, it would have been additive to any stray voltage induced through the CA42282 pylon electrical disconnect and could have acted as a trigger if the simultaneous malfunctions had occurred, as previously described in Opinion 5.

14. That the effects of high energy radiation on the LAU-10/A's from the ALQ-51, installed in A-4s in close proximity to F-4 #110, were not known, but are considered not to be the proximate cause of firing the ZUNI.

15. That the various armament safety features in the F-4B aircraft, the TER-7 and the LAU-10/A-ZUNI combination, if functioning entirely properly and properly used, provide sufficient safety for attack carrier operations. This assumes that all defects in hardware and deficiencies in procedures noted herein have been corrected.

- -

16. That the operational and technical procedures developed by FORRESTAL and the embarked air wing for operations on Yankee Station relating to ordnance handling were poorly documented and promulgated.

17. That approved procedures should have been adhered to, even during conditions of high tempo operations. That if normal procedures were not found adequate, they should have been officially modified to fit combat operational requirements.

18. That documentation by VF-11 regarding detailed ordnance procedures was found to be inadequate.

19. That it was basically a sound decision to organize composite air wing catapult arming crews in order to reduce the numbers of personnel in the vicinity of the catapults.

20. That the organization and specific duties of each member of the composite air wing arming crews should have been detailed in writing and promulgated over the Carrier Air Wing Commander's signature.

21. That on 29 June 1967 the augmented Weapons Coordination Board discussed and made decisions concerning ordnance handling and safety procedures on board FORRESTAL. Those decisions falling within his authority should have been approved by the Commanding Officer, FORRESTAL, and promulgated in writing to all departments, divisions and squadrons concerned over his signature. Where deviations from documented safety or other procedures were involved, approval should have been specifically requested from authority higher than CO, FORRESTAL, to deviate.

22. That if rockets were to have been plugged in prior to reaching the catapults, the procedures therefor should have been approved and documented by authority higher than the CO, FORRESTAL.

23. That the minutes of the 29 June meeting should not have been informally distributed to the CVW-17 Ordnance Officer as a double-spaced rough but should instead have been officially promulgated to squadron commanding officers and ship's department heads over the Operations Officer's signature.

24. That the reply to the double-spaced rough minutes of the 29 June meeting, which reply was originated by AOC b-b should not have gone directly to the CVW-17 Ordnance Officer without having been reviewed by an appropriate officer in VF-11. The reply should have gone through the squadron chain of command for review (even though the rough minutes were informally received) and should have been forwarded over the CO, VF-11's signature.

25. That LTJG b-b by not referring the above double-spaced rough to his commanding officer, failed to keep his CO properly informed of matters relating to ordnance procedures.

26. That CDR b-b was aware of the 29 June meeting and its possible impact on squadron ordnance operations, but failed to keep himself adequately informed of the details relating thereto.

27. That the ordnancemen of VF-11 were generally competent as individuals but were poorly organized and instructed.

28. That official documentation as to when and where to conduct stray voltage checks for LAU-10/As is ambiguous and inadequate.

29. That documentation as to specifically how to conduct stray voltage checks for the F-4B/TER-7/LAU-10/A weapons system is inadequate.

30. That although there were no directives or guidance to prohibit it, leading VF-11 ordnance personnel exercised poor judgment in allowing stray voltage checks and plugging in rockets before the aircraft's electrical system had stabilized after starting, i.e., before both engines were started and the aircraft was switched to internal electrical power.

31. That ʰᵇ action in conducting stray voltage checks on F-4 #110 before both engines had been started and the aircraft electrical power had stabilized was contrary to VF-11 squadron policy, as stated by the CO. However, this policy was not set forth in writing.

32. That ʰ-ᵇ and AO ʰ-ᵇ VF-11, acting on incomplete squadron instructions gave incorrect instructions to ɜ-ᵇ as to the time when stray voltage checks should be started.

33. That ʰ-ᵇ arried out the stray voltage checks on F-4 #110 on 29 July completely in accordance with the instructions which had been given him.

34. That the ordnance crew which was working on F-4 #110 was loosely organized and without adequate supervision.

35. That the testimony of ʰ-ᵇ is false in that he was not standing, as he testified, directly in front of the port LAU-10/As of F-4 #110 at the time the fire was initiated.

36. That the testimony of 🖎🖎 and 🖎🖎 is false in regard to their stated failure to observe the ZUNI rocket fire from #110.

37. That the sworn statement made by 🖎🖎 on 1 August wherein he stated that he did see the ZUNI rocket fire, is true.

38. That if 🖎🖎 were standing in the position he stated, he must have seen the ZUNI rocket fire and therefore his testimony is either false in regard to his location or false with regard to failure to see the ZUNI.

39. That the testimony of LCDR 🖎🖎 to the effect that ZUNI missile covers were intact, applies, in the opinion of the Board, to missile covers on the starboard side of the aircraft. That LCDR 🖎🖎 did not accurately inspect the port side ZUNI missiles.

40. That the CO, VF-11, displayed poor judgment in eliminating the requirement for an ordnance safety petty officer during stray voltage check and rocket plug-in while aircraft were spotted in the pack.

41. That ordnancemen should not conduct stray voltage checks or plug in rockets without the pilot's specific knowledge. Hence, CO, VF-11 erred in abolishing a hand signal exchange in the pack with the pilot that such ordnance functions were being conducted.

42. That the preflight briefing given by CDI 🖎🖎 to LCDR 🖎🖎 was inadequate in that more attention should have been directed to flight deck inspection procedures relating to the LAU-10/A ZUNI combination which LCDR 🖎🖎 had not previously employed.

43. That VF-11 preflight procedures are deficient in that they should have required the pilot (and RIO) to check the position of the LAU-10/A shorting device prior to manning the aircraft.

44. That NATOPS procedures are deficient in that they do not require that the pilot check the position of the Armament Override Switch during pre-starting cockpit check.

45. That the act of actuating the Home-Step switch on the TER-7 after the LAU-10/As had been plugged in was a potentially dangerous action in that if he stepped rather than homed the switch, he might have fired a LAU-10/A if certain malfunctions (Opinion 5) existed in the firing circuit. There were, however, no written or verbal prohibitions to this action in effect on 29 July.

46. That the situation described in the preceding opinion is further aggravated by reversal of the direction of throw of the Home-Step switch with the issuance of the -527 model TER-7.

47. That the personnel assigned to the Weapons Branch of VF-11 were adequate in both numbers and experience levels, and lack of personnel or experience level was not a contributing cause of the accident.

48. That VF-11 had sufficient time and opportunity since transitioning to F-4 aircraft in July 1966 to develop operational proficiency in the F-4. The squadron's administrative and organizational progress is unimpressive.

49. That the loose ordnance organization and poor procedures of VF-11 as described in previous statements of opinion were not known to the Air Wing Commander, the Air Wing Ordnance Officer, the Ship's Operations Officer or CO, FORRESTAL.

50. That inconsistencies and improper procedures evident in VF-11 ordnance operations should have been discerned by ship and air wing personnel and corrected.

51. That, at least a part of the poor organization and procedures, mentioned above, and the failure to uncover them can be attributed to the short period during which the squadron had been operating on Yankee Station.

52. That despite the short interval between the ship's overhaul and deployment to WESTPAC and despite the relatively short period since commissioning of CVW-17, FORRESTAL arrived on Yankee Station with a comparatively high degree of personnel and material readiness.

53. That Navy personnel policies should have permitted the stabilization of the ship's personnel at the beginning of refresher training and thus have enabled FORRESTAL to retain, for deployment, those personnel who received refresher training.

- -

54. That FORRESTAL's material readiness for fire fighting and damage control were at acceptable standards at the time of the fire.

55. That the magnitude of the fire and the resultant heavy damage was due to the concentration on the flight deck of aircraft loaded with aviation ordnance stores and huge quantities of aviation fuel; a condition characteristic of present day combat carrier operations.

56. That with existing installed fire fighting equipment, the fire could not have been extinguished prior to the explosion of major ordnance (94 seconds after initiation of the fire) regardless of the aggressiveness, readiness, response and expertise of personnel and readiness of equipment.

57. That extensive fire fighting efforts were underway on the flight deck at the instant of the first major explosion.

58. That effective fire fighting of this large fire could not have been conducted during the period of the major explosions (approximately five minutes) and was properly suspended for this period until after major explosions had subsided.

59. That the design and operating procedures of fire fighting equipment currently available in attack carriers is totally inadequate to the needs generated by modern combat operations and the concentrations of very large quantities of ordnance and fuel on jet aircraft.

112

60. That because of the tremendous quantities of fuel carried by jet aircraft, which may be expected to spill to the deck in casualties of this type, methods and devices must be developed to rapidly jettison or drain over the side large quantities of fuel which may flow onto the deck.

61. That, though not a significant factor to the spread of the fire in this instance, difficulty was experienced in jettisoning ordnance stores and aircraft. The RA-5C was particularly difficult to jettison because of its size and weight.

62. That the fire in FORRESTAL might have been confined to a few aircraft had proper equipment and techniques been available to either rapidly jettison ordnance stores on burning aircraft, or, to cool the stores while the fire was being extinguished.

63. That the concentration during combat operations of exposed ordnance stores on the hangar and flight decks in so-called "bomb farms" which have neither special fire fighting protection nor high capacity emergency jettison facilities creates a dangerous situation.

64. That on the forenoon of 29 July 1967, when large quantities of ordnance stores were exposed above the second deck, the manning of only four HCFF generating stations on the second deck was insufficient to provide required protection against fire.

65. That personnel not familiar with the functions of damage control during general quarters, improperly initiated many actions, such as opening and closing hangar bay doors without informing Damage Control Central.

66. That concentrations of armed aircraft on the flight deck generate high hazard conditions for the flight deck as well as for the 03, 02, 01 and hangar deck areas beneath. In such circumstances, concentrations of personnel on decks below the affected part of the flight deck are highly hazarded.

67. That cook-off times of ordnance stores in use were not available to FORRESTAL and that considerable injury and loss of life can be attributed to the cook-off of installed ordnance stores at a time earlier than expected.

68. That the present HCFF hose stations in FORRESTAL lack standard configuration and invite confusion, and therefore constitute a potentially hazardous situation.

69. That the technique for perforating a deck or bulkhead in order to insert a fire hose into a compartment that is otherwise unreachable has great merit. However, the directional characteristics of currently available hose nozzles, which cannot be manipulated through the hole to cover all areas within a compartment, limit the effectiveness of this technique.

70. That key personnel of the damage control organization, including key repair party members, were not distinctively marked and could not be readily identified. In some instances this created confusion concerning leadership, control or even the presence of damage control personnel.

71. That FORRESTAL's allowance of OBA cannisters and fog foam was sufficient to support only the initial fire fighting efforts. It was insufficient to support the sustained effort required by this major fire.

72. That without the continuing replenishment of OBA cannisters and fog foam from assisting ships, FORRESTAL would probably not have been able to extinguish all fires but could probably have contained them until they burned out.

73. That, although FORRESTAL was 25 OBAs and 50 cans of fog foam short of allowance at initiation of the fire the shortage of these items did not inhibit initial damage control and fire fighting efforts.

74. That although at allowance, insufficient numbers of eductors and portable blowers (red devils) were on hand to adequately cope with clean up operations during the terminal periods of the emergency.

75. That FORRESTAL's fire fighting operations required the rapid uncoupling and recoupling of fire hoses. Presently available couplings and fittings are too cumbersome for the instantaneous response now required.

76. That aircrewmen are not adequately trained in the considerations and techniques for abandoning static aircraft engulfed in flames.

77. That current fire fighting exercises do not provide adequate training for the type and scope of fire experienced by FORRESTAL 29 July.

78. That the 1MC ship's general announcing system as presently installed in FORRESTAL is inadequate for passing the word effectively on the hangar deck.

79. That the 750 gallons of liquid oxygen stored in compartment 1-192-2-E adjacent to hangar bay 3 constituted a high hazard for a prolonged period (approximately one hour) while the liquid oxygen was being drained over the side through a stall hose.

80. That the weather was not a contributing factor to the casualty although the high ambient temperature may have reduced the cook-off time of ordnance stores. The hot decks made the spilled JP-5 fuel more volatile and could have been an additional factor in the rapid spreading of the fire.

81. That the divisional doors between hangar bays 2 and 3 were closed and the sprinkler system in hangar bay 3 actuated early enough to effectively prevent the spread of fire forward in hangar bay 3 where ten aircraft were spotted.

82. That fire boundaries on the flight deck and below were initially established at the optimum locations considering prevailing conditions.

83. That some hatches and doors along casualty routes for injured personnel properly remained open during condition ZEBRA to enhance the movement of injured personnel.

84. That the status and loading of each magazine in the ship is vital information for effective damage control. This information should be continuously available in Central Control.

85. That records were not properly maintained and preserved in Central Control and Damage Control Central relating to significant events that occurred during the casualty.

86. That the FORRESTAL electrical systems were managed in a highly satisfactory manner throughout the entire period of the emergency.

87. That during the fire, various untrained individuals took well-intentioned but ill-advised actions because they were unaware that damage control personnel were at the scene and were executing a considered plan of action.

88. That air wing personnel require considerably more basic training in fire fighting and damage control.

89. That good judgment was exercised in sounding general quarters at 1053H, 29 July 1967, as soon as the magnitude of the fire became apparent.

90. That the Commanding Officer FORRESTAL consistently demonstrated a personal interest in the material condition and training of his ship's company in fire fighting and damage control.

- -

91. That Captain   performed satisfactorily his assigned duties as Commanding Officer, USS FORRESTAL (CVA-59), and that no blame attaches to Captain   in connection with the fire that occurred in FORRESTAL on 29 July 1967.

92. That Commander   performed satisfactorily his assigned duties as Engineering Officer, USS FORRESTAL (CVA-59), and that no blame attaches to Commander   in connection with the fire that occurred in FORRESTAL on 29 July 1967.

93. That Lieutenant Commander   performed satisfactorily his assigned duties as Damage Control Assistant, USS FORRESTAL (CVA-59), and that no blame attaches to Lieutenant Commander   n connection with the fire that occurred in FORRESTAL on 29 July 1967.

94. That Commander   performed satisfactorily his assigned duties as Air Officer, USS FORRESTAL (CVA-59), and that no blame attaches to Commander   in connection with the fire that occurred in FORRESTAL on 29 July 1967.

95. That Lieutenant Commander   performed satisfactorily his assigned duties as Hangar Deck Officer, USS FORRESTAL (CVA-59), before and after the fire. Lieutenant Commander   as not on board FORRESTAL 29 July during the fire. No blame attaches to Lieutenant Commander   in connection with the fire that occurred in FORRESTAL on 29 July 1967.

96. That Lieutenant   performed satisfactorily his assigned duties as Flight Deck Officer, USS FORRESTAL (CVA-59), and that no blame attaches to Lieutenan   n connection with the fire that occurred in FORRESTAL on 29 July 1967.

All redactions are D-6

117

97. That Chief Warrant Officer ___ ⸝rformed satisfactorily his
assigned duties as Fire Marshall, USS FORRESTAL (CVA-59), and that
no blame attaches to Chief Warrant Officer ___ n connection with
the fire that occurred in FORRESTAL on 29 July 1967.

98. That Warrant Officer ___ performed satisfactorily his assigned
duties as Air Ordnance Gunner, USS FORRESTAL (CVA-59), and that no
blame attaches to Warrant Officer ___ in connection with the fire that
occurred in FORRESTAL on 29 July 1967.

99. That Commander ___, Jr., performed satisfactorily his assigned
duties as Commander, CVW-17 and that no blame attaches to CDR ___ in
connection with the fire that occurred in FORRESTAL on 29 July 1967.

100. That Lieutenant Commander ___ performed satisfactorily his
assigned duties as pilot of VF-11 aircraft #110 and that no blame
attaches to Lieutenant Commander ___ n connection with the fire
that occurred in FORRESTAL on 29 July 1967.

'101. That because of overall inherent weaknesses in the development and
documentation of ordnance operations, procedures and safety measures,
Lieutenant ___, CVW Ordnance Officer, was unable to carry out all
of his responsibilities relating to safety and ordnance handling
operations in an efficient manner, however, no blame attaches to
Lieutenant ___ n connection with the fire that occurred in
FORRESTAL on 29 July 1967.

102. That CDR ___ demonstrated poor judgment and lack of supervision in the following areas:

    a. Eliminating the "hands off" signal and safety petty officer
during ordnance evolutions in the pack.

    b. Permitting loose organization and operational procedures to
continue in the Weapons Branch of VF-11.

    c. Allowing squadron instructions and organization manual to
become outdated.

However, no blame attaches to CDR ___ n connection with the fire
that occurred in FORRESTAL on 29 July 1967.

113

103. That Lieutenant Commander   performed satisfactorily his assigned duties as Maintenance Officer, VF-11, and that no blame attaches to Lieutenant Commander   in connection with the fire that occurred in FORRESTAL on 29 July 1967.

104. That Lieutenant   performed satisfactorily his assigned duties as Avionics/Weapons Officer, VF-11, and that no blame attaches to Lieutenant   in connection with the fire that occurred in FORRESTAL on 29 July 1967.

105. That Lieutenant (junior grade)   did not exercise the required close degree of supervision and control over personnel in the ordnance branch of VF-11, however, no blame attaches to Lieutenant (junior grade)   n connection with the fire that occurred in FORRESTAL on 29 July 1967.

106. That Warrant Officer   performed satisfactorily his assigned duties as Assistant Avionics/Weapons Officer, VF-11, and that no blame attaches to Warrant Officer   n connection with the fire that occurred in FORRESTAL on 29 July 1967.

107. That AOC   satisfactorily performed his duties as Aviation Ordnance Supervisor, VF-11, and no blame attaches to AOC   in connection with the fire that occurred in FORRESTAL on 29 July 1967.

108. That AO1   performed satisfactorily his assigned duties as aviation ordnance team leader, VF-11, and that no blame attaches to AO.   in connection with the fire that occurred in FORRESTAL on 29 July 1967.

109. That AO2   VF-11, satisfactorily performed the ordnance evolutions assigned him at F-4 #110 and that no blame attaches to AO2   nnection with the fire that occurred in FORRESTAL on 29 July 1967.

All redactions are B-6.

119

110. That AOAM _____ VF-11, performed satisfactorily his assigned duties as stray voltage checker in accordance with instructions given him and that no blame attaches to AOAM _____ in connection with the fire that occurred in FORRESTAL on 29 July 1967.

- -

111. That the major cause of injuries were wounds caused by shrapnel, flying objects and burns.

112. That the major causes of death were concussions from the bombs, suffocation, burns, and wounds caused by shrapnel and flying debris.

113. That all personnel injured during the fire on board FORRESTAL on 29 July 1967 were injured in the line of duty and not due to their own misconduct.

114. That the deaths and injuries resulting from the fire aboard FORRESTAL on 29 July 1967 were not caused by the intent of any person or persons in the naval service or connected therewith.

115. That the deaths and injuries resulting from the fire aboard FORRESTAL on 29 July 1967 were not caused by the intent, fault, negligence, or inefficiency of any person or persons embarked in FORRESTAL.

116.

B-5

RECOMMENDATIONS

1. That the LAU-10/A launcher not be used on F-4 aircraft until:

    a. A redesigned shorting device is available.

    b. Deficiencies in CA42282 pylon electrical disconnect are corrected.

    c. Wiring of the -527 TER safety switch is corrected.

    d. Adequate procedures are promulgated to check for stray voltage in the rocket firing circuits of the F-4.

2. That the LAU-10/A be provided, as a matter of urgency, with a shorting device which acts positively, assuring that it can be placed only in "safe" or "arm" with no intermediate positions.

3. That the above LAU-10/A shorting device be so designed that it does not interfere with the LAU-10/A suspension lug.

4. That positive effective action be instituted to correct electrical shorting discrepancies in the CA42282 pylon electrical disconnect of F-4 aircraft, as a matter of urgency.

5. That the wiring of the safety switch in the -527 TER be changed so that it opens or severs the firing circuit when in the "SAFE" position (as in the -505 and -521 TERs), rather than be shunted to ground.

6. That a stray voltage test receptacle be incorporated in the TER-7 in order to eliminate the requirement for home-made adapter cables in checking for stray voltage in the rocket firing circuits.

7. That procedures for conducting firing circuit continuity checks on the TER-7 MOD-527 be immediately modified to prevent damage to rocket firing circuitry.

8. That care be taken during the design of new or modified equipment to insure that the direction of actuation of the same control, or of controls which perform the same function, is not changed. Enclosure (368) shows, as an example, the change in direction of actuation of the HOME-STEP switch on the TER-7 between models -505 and -521 and, model -527.

9. That directives from appropriate authority be issued specifically stating when and where rocket firing circuit stray voltage checks and rocket launcher plug-ins will be accomplished under the various conditions of operation; (land, based, carrier based) considering various weapon system combinations.

10. That directives from appropriate authority be issued which specifically state the precise procedures for conducting rocket firing circuit stray voltage checks prior to rocket plug-in.

11. That an immediate technical review of all aircraft/rocket systems, including the ZUNI, be made to verify and correct procedures and documentation for their use, and that existing supporting documentation be corrected and issued as a matter of urgency.

12. That each USN aircraft weapon system (the basic aircraft and its combinations of ordnance equipments and ordnance stores) be thoroughly reviewed by competent experts to develop, verify or correct all pertinent procedures and safety precautions. That the results thereof be appropriately issued.

13. That electronic equipment which will add to the electronic environment of carriers not be sent to the fleet until its RADHAZ effects have been analyzed and tested with all ordnance currently in use.

- -

14. That operational procedures developed by attack carrier forces during S. E. Asia operations be fully documented and distributed by CNO to all commands concerned.

15. That any procedures so developed which are in conflict with standard Navy safety precautions be given the closest possible scrutiny by experts so that a decision can be made as to whether the advantages to be gained are in fact worth any added risks involved.

16. That added emphasis be placed on adherence to approved procedures, even under conditions of high tempo combat operations.

17. That if approved procedures are not found adequate for combat situations, the affected command initiate immediate recommendations for specific modification of procedures to appropriate higher authority.

18. That the CNO establish a single series of publications similiar to or a part of the NWP (Naval Warfare Publication) series which will serve as a single source of documentation for all operational and technical procedures related to the operation of aircraft and appertaining systems. This series of aircraft operational and technical procedures (as differentiated from maintenance (3M) publications) should be kept under continual review by knowledgeable experts.

19. That the above series of publications be promulgated by a single coordinating office to assure the completeness, standardization, continuity and interface of the aircraft publications.

20. That the above series of publications incorporate a system for frequent periodic review and updating, incorporating recommendations from the operating forces as well as from the technical commands.

21. That CO FORRESTAL develop and use a standardized procedure for preparing and documenting all operational and technical procedures not adequately covered by other official publications.

22. That each squadron of CVW-17 provide the Air Wing Commander in writing with the precise procedures and safety measures the squadron will follow in all actions related to weapons and ordnance.

23. That the Air Wing Commander coordinate, standardize and recommend improvements to the procedures submitted in accordance with the preceding paragraph. When satisfied that the procedures are safe, efficient and in compliance with directives, and following the ship CO's approval, when embarked, he then give written approval of the specific procedures.

123

24. That CO, VF-11, immediately prepare and issue appropriate instructions as to the organization, and, ordnance operating and safety procedures for the squadron.

25. That COMCVW-17 set out in writing the organization, function and duties of all composite air wing organizations such as the composite catapult arming crew.

26. That CO, VF-11 issue appropriate directives specifically stating the requirement that an air ordnance safety petty officer be designated to supervise personnel at all times that live ordnance is being handled.

27. That, in recognition of the special importance of safety in ordnance handling evolutions, VF-11 squadron organizational instructions be revised to require the Weapons Officer to report directly to the Commanding Officer for matters relating to ordnance handling safety.

28. That VF-11 ordnance teams be composed of personnel who are specifically designated by name; that such teams should invariably train and operate as integral units; that the senior petty officer of each team be designated as the team leader; and that the team leader remain with the team, actively supervising operations in progress. If the team must be broken up into smaller units, then the senior petty officer of each unit must function as the unit leader, remaining with the unit and actively supervising operations in progress.

29. That after a pilot has preflighted and manned his aircraft, the aircraft should not thereafter be touched without informing the pilot, by appropriate hand signals, of the nature of the function intended to be accomplished and when such function has been completed.

- -

30. That increased emphasis be exerted by the Chief of Naval Personnel to minimize the detachment of trained personnel from ships and squadrons in the last few months prior to deployment.

- -

31. That a study be initiated to develop revolutionary new fire fighting equipment and procedures which are responsive to existing requirements of instantaneous reaction.

32. That carrier commanding officers use extraordinary measures and ingenuity in pre-positioning and readying available fire fighting resources, both equipment and personnel, to obtain the quickest, most effective fire fighting capability that is practicable.

33. That an immediate study be instituted to recommend modifications to existing carrier fire fighting systems to increase response and effectiveness, including the following:

    a. Deck edge HCFF monitors served by existing risers.

    b. HCFF stations to be individually actuated either remotely from Pri Fly or from the catwalk.

    c. A system of water curtains or sprinklers to rapidly wash large quantities of aircraft fuel off the deck and over the side.

    d. A system of scuppers and drains capable of draining off large volumes of water and fuel. These scuppers and drains should be so designed as to minimize spreading of fuel and fire to lower decks, sponsons and fantail.

    e. HCFF and salt water monitors on the island, capable of issuing large volumes of foam and water.

    f. Current HCFF generators be re-engineered to provide foam to the flight and hangar deck in less than five seconds, if practicable.

34. That a more effective method of storing HCFF and salt water fire hose on the flight deck be devised to facilitate leading out hose and eliminate tangling and fouling.

35. That controls for HCFF hose stations be of a standard configuration, with distinctive markings, isolated from other fittings which might be confusing. The stations' location should be highly visible and unmistakably identifiable from the flight and hangar deck by a standard method which readily attracts attention and is not obscured by aircraft spotted nearby.

36. That key personnel, particularly key repair party members, be issued and wear more distinctive badge/hat/brassard which is readily discernible to promote better on-the-scene control and identification.

37. That Central Control in FORRESTAL maintain daily records as to the status and loading of each magazine.

38. That FORRESTAL allowance of fire fighting equipment be modified as follows:

ITEM	CURRENT ALLOWANCE	RECOMMENDED ALLOWANCE
Fog foam (5 gal)	1220	2500
OBA	550	620
OBA cannisters	3300	8000

39. That deficiencies in the 1MC carrier's general announcing system be corrected to provide intelligible transmission of messages to personnel in the hangar bays.

40. That when practicable fire lanes be created in the pack by leaving space between aircraft in order to assist in isolating fires and to allow hoses and equipment to be moved with greater facility.

41.   That procedures and equipment be developed for rapidly jettisoning ordnance stores that are concentrated in "bomb farms" on the hangar or flight deck preparatory to loading on aircraft.

42.   That additional jettison ramps, including one located outboard of the island, be installed on carriers to facilitate jettisoning of ordnance and dangerous equipment, e.g., LOX carts.

43.   That equipment and procedures be developed to rapidly jettison burning aircraft; for example, bulldozer attachments for yellow vehicles.

44.   That portable aircraft jettison ramps be provided so that heavy aircraft may be pushed over the side at any location on the flight deck without hanging up in the catwalk.

45.   That procedures be developed and that ordnancemen and fire fighters be trained to accomplish rapid dearming of aircraft threatened by fire.

46.   That a list of the cook-off times of each appropriate ordnance store be printed on a decal to be posted at fire fighting stations and on appropriate fire fighting equipment afloat and ashore.

47.   That each item of ordnance be labelled with the cook-off time for that particular store.

48.   That equipment and procedures be developed to prevent cook-off of ordnance stores while fire fighting is in progress on an aircraft, for example:

a.   A water fed muff or cone to cover the stores with cooling water.

b.   Spray or water curtain rigs which can be rapidly applied to ordnance stores.

49. That consideration be given in future carrier modification and new designs to extending the flight deck over the stern to prevent burning fuel from engulfing the fantail.

50. That consideration be given to providing all appropriate compartments (particularly berthing, living and working) with alternate escape exits.

51. That remotely actuated, high capacity (pop-up) sprinkler systems be considered for installation to cover the entire carrier flight decks.

52. That remotely actuated high capacity spray or water curtain systems be installed on attack carriers to provide immediate fire protection in those specific areas on the hangar deck and flight deck where exposed ordnance stores may be concentrated preparatory to being loaded on aircraft.

53. That consideration be given to employing an armored fire fighting vehicle on the flight deck of attack carriers engaged in combat operations which:

    a. Is capable of extinguishing aircraft fires including fuel, oxygen, and various types of explosive ordnance.

    b. Is able to bulldoze burning wreckage from deck.

    c. Protects operators from shrapnel and from ordnance detonations.

    d. Is able to continue to operate in an environment of fire and explosions of ordnance;

    e. Has two-way radio communication with Pri Fly.

54. That a special purpose omni-directional nozzle be developed and issued, possibly similar to a Butterworthing nozzle, which can be inserted through a small hole and will spray all areas within a compartment.

55. That tape recorders capable of simultaneously recording all communications with Central Control/Damage Control Central be installed on carriers, as a training aid and to assist in investigating casualties. Each sound powered circuit should be on a separate channel.

56. That specific standardized procedures be established for recording data on important events in Damage Control Central/Central Control during a casualty.

57. That Carrier Air Wing personnel receive more formal training in fire fighting and damage control with specific emphasis on:

a. Fundamentals of ship's damage control organization and operations.

b. Principles involved in controlling damage aboard ship.

c. Basic knowledge of ship's geography, including flow patterns and escape routes.

d. Fundamentals and use of basic damage control and fire fighting equipment with particular emphasis on OBA, fog foam, firemain and sprinkler systems and how to actuate and use them.

58. That specific minimum qualifications in damage control and fire fighting be established, that all personnel assigned to carriers (including air wing personnel) meet these qualifications prior to embarking, and that satisfactory completion of these qualifications be made a matter of record in each man's jacket.

59. That flight deck fire fighting exercises be developed to train personnel in fighting fires of the type experienced by FORRESTAL, taking into consideration the fire's magnitude, the live ordnance and early casualties to key personnel and equipment.

60. That techniques be developed for abandoning aircraft which are engulfed in flames, to provide aircrewmen with the most effective method of escape. Many variables should be considered such as escape time available, relative wind, possible abandonment and escape routes, size and type of fire, stores carried, whether to secure engines, taxi clear, eject, etc.

61. That the attached "Lessons Learned" be promulgated to appropriate commands at earliest concurrence by convening authority; in advance of routine routing of the record of this Board of Investigation.

- -

62. That no disciplinary or administrative action be taken with regard to any persons attached to USS FORRESTAL (CVA-59) or Carrier Air Wing 17 as a result of the fire which occurred on board USS FORRESTAL on 29 July 1967.

Rear Admiral, U. S. Navy

Captain, U. S. Navy

Captain, U. S. Navy

Final Entry.

Rear Admiral, U. S. Navy

Commander, U. S. Navy
Counsel for the Board

Lieutenant Commander, U. S. Navy   B-6
Assistant Counsel for the Board

All redactions are

130